Rare and Endangered Species of Bruce and Grey Counties, Ontario

A Guide to Provincially Rare, Vulnerable, Threatened and Endangered Species Found in Bruce and Grey

The Bruce-Grey Plant Committee
Owen Sound Field Naturalists

Stan Brown Printers Limited
Owen Sound, Ontario

Canadian Cataloguing in Publication Data

Main entry under title:
Rare and Endangered Species of Bruce and Grey Counties

Includes bibliographical references and index.
ISBN 0-9680279-4-6

1. Rare animals—Ontario—Grey (County) 2. Rare plants—Ontario—Grey (County) 3. Rare animals—Ontario—Bruce (County) 4. Rare plants—Ontario—Bruce (County). I. Owen Sound Field Naturalists. Bruce-Grey Plant Committee

QH106.2.O5R365 2001 578.68'09713'18 C2001-900521-0

Books available from The Bruce-Grey Plant Committee
A Checklist of Vascular Plants for Bruce and Grey Counties, Ontario – $5.00
A Guide to the Orchids of Bruce and Grey Counties, Ontario – $15.00
A Guide to the Ferns of Grey and Bruce Counties, Ontario – $15.00
The Asters, Goldenrods and Fleabanes of Grey and Bruce Counties, Ontario – $8.00

For mail orders please add postage.

All books available from:
The Bruce-Grey Plant Committee
c/o The Owen Sound Field Naturalists
Box 401, Owen Sound, Ontario N4K 5P7

Fern Guides may also be obtained from:
The Toronto Field Naturalists
2 Carlton Street, No. 1519
Toronto, Ontario M5B 1J3
Telephone: 416-593-2656
Office open Fridays 9:00 a.m. to 12:00 noon

Table of Contents

N
Fathom Five National Marine Park
TOBERMORY
CABOT HEAD
Bruce Peninsula Nat. Park
Dorcas Bay
DYERS BAY
Johnston Harbour
MILLER LAKE
CAPE CHIN
Bruce Trail
GEORGIAN BAY
STOKES BAY
LIONS HEAD
FERNDALE
6
HOPE BAY
Cape Croker I.R. #27
PIKE BAY
LAKE HURON
RED BAY
MAR
Colpoy's Bay
Big Bay
Isaac L.
WIARTON
OLIPHANT
OWEN SOUND BAY
Rankin R.
Chiefs Pt. I.R. #28
Meaford Tank Range
SHALLOW LAKE
SAUBLE BEACH
LEITH
MEAFORD
26
Sauble R.
THORNBURY
OWEN SOUND
Bognor Marsh
21
Saugeen I.R. #29
SOUTHAMPTON
Bruce Trail
TARA
6/10
CRAIGLEITH
KEADY
CHATSWORTH
PORT ELGIN
13
MacGregor Point
HOLLAND CENTRE
Baie du Doré
KIMBERLEY
BNPD
WILLIAMSFORD
10
Inverhuron Prov. Park
PAISLEY
MARKDALE
4
CHESLEY
FLESHERTON
BRUCE
GREY
KINCARDINE
Saugeen R.
HANOVER
DURHAM
DUNDALK
Greenock Swamp
WALKERTON
9
RIPLEY
VARNEY
NEUSTADT
HOLYROOD
MILDMAY
AYTON
TEESWATER
MOUNT FOREST
89
86
LUCKNOW

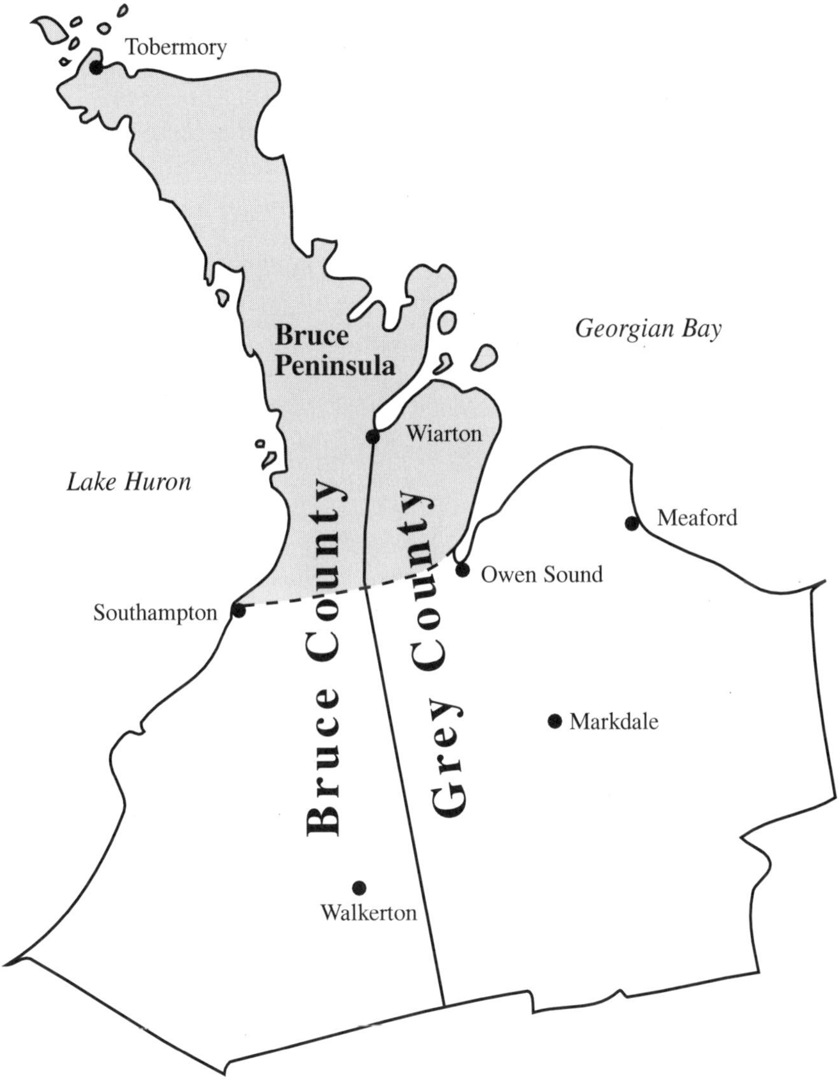
Tobermory
Georgian Bay
Bruce Peninsula
Wiarton
Lake Huron
Meaford
Owen Sound
Southampton
Bruce County
Grey County
Markdale
Walkerton

Preface

The idea for this book originated with Bob Gray, Area Ecologist for the Ministry of Natural Resources in Owen Sound. He started producing fact sheets for rare, threatened and endangered species for use in encouraging landowners to preserve natural habitat. Changes in funding and organization prevented completion of this project. The Bruce-Grey Plant Committee decided that the material should be used to produce a book which would be available to local residents, naturalists, students, land use planners and landowners. Other members of the Owen Sound Field Naturalists were recruited to deal with the fauna. In addition, many other organizations and individuals have been most helpful. These contributions are acknowledged below. No funding was received from any other organization. It is hoped that this publication will raise awareness of the unique qualities of this area and the importance of maintaining healthy ecosystems. It should be noted that, with a few exceptions, only provincially rare native species listed by the Natural Heritage Information Centre (NHIC) at S1 to S3 in Ontario are included. Non-native species which have been introduced have not been considered. Locally or regionally rare species also have not been discussed in this publication. Birds presented particular problems due to their mobility and rapid response to climate change, as well as difficulties with reliable identification. Choices as to which species should be included had to be made. Much work still needs to be done to determine provincial rarity for species of lower plants and animals such as mosses, insects and molluscs. No amphibians have been ranked as provincially rare for this area but they are, undoubtedly, under considerable pressure as the viablity of wetlands is increasingly threatened. Many of the lower groups also present considerable technical difficulties in identification. For these reasons these groups were considered to be beyond the scope of this book, but it is recognized that they are all an important part of ecosystems and often very good indicators of environmental health.

Acknowledgements

Members of the Bruce-Grey Plant Committee:

Mac Campbell — for assisting with proof reading and editing and for providing information on southern Bruce County.

Joan Crowe — project coordinator, also for writing parts of text, proof reading and editing.

Bob Gray — for initiating this project with the production of fact sheets on rare species and for making this material and illustrations available for this project. For expert advice on local ecology and assistance with researching material and for many helpful suggestions.

Ellen and Orris Hull — for assisting with research, writing, proof reading and editing.
Joe Johnson — for expert advice on plants and other species, especially with respect to local distribution and local rarity, for writing local distribution and habitat for plants, and for reviewing and proof reading the entire manuscript.
Nels Maher — for information on ferns and other species, for proof reading, mapping, and expert advice on, and assistance with, printing and publishing.
Assistance also from:
Walter Crowe — for researching and writing the preliminary text for the birds. Also for proof reading and computer assistance.
Sandy Dobbyn: — for writing the mammal section and for proof reading.
Dave Fidler — for local distribution information on breeding birds.
Erling Holm — (Curatorial Assistant Icthyology and Herpetology, Royal Ontario Museum) for information on fish species and for proof reading and making amendments to the text.
Daphne Johnston — Cape Croker librarian for giving access to the Cape Croker library and making available information about the Missisauga-Ojibwa people of this area.
Phil Kor, Senior Conservation Geologist, Ontario Parks — for input on geology.
Carol-Ann Lacroix — University of Guelph Herbarium for the loan of specimens.
Jean Maher — for assisting with proof reading.
Andy McKee, area biologist — for producing information on fish and checking the text.
Peter Middleton — for local distribution information on breeding birds.
Michael Oldham — for making available information on rarity status from the National Heritage Information Centre and for proof reading the manuscript and making many helpful suggestions, especially with respect to reptiles and plants.
George Peck — for overseeing and writing the bird text and for providing information on local distribution of nesting birds. Also for proof reading and editing.
Lynne Richardson — for reviewing the bird section of the manuscript and for valued suggestions.
Joan Venn — University of Waterloo Herbarium for the loan of specimens.
Illustrations:
Dr. E.J. Crossman and Dr. W.B. Scott — for generously allowing us to use diagrams from their book Freshwater Fishes of Canada. 1998.
Paul Harpley — for allowing us to use his original drawings made for the book Familiar Amphibians and Reptiles of Ontario.
Oxford University Press — Ann Checchia Rights and Permissions Editor. For permission to use the illustration of the Eastern Small-footed Bat from the Mammals of Eastern Canada by Randolph Peterson.
George Peck — for providing the bird photographs.
The Staff at Stan Brown Printers — for their expert assistance and excellent quality of work.

Introduction

The words **rare, vulnerable, threatened** and **endangered** may be applied to many different species and may be defined in many different ways. However, for a species to be officially ranked as rare there are two criteria; first geographic (restricted area), second demographic (low numbers), both of which vary considerably. For a species to qualify as rare it must have the correct mix of these two factors.

Rarity can occur for a variety of reasons and all of these are probably applicable to Bruce and Grey counties. When environmental conditions change for whatever reason, certain species may be unable to adapt to the new regime and will gradually disappear. This has happened innumerable times in the course of evolution. At the same time rare genetic variations of that species may respond to the changing environment and become established as a new species. At a given point in time both species will be rare, but one will be on the verge of extinction and the other on the verge of expansion. Extinction is a natural part of the evolution of species. Another cause of rarity may be that the species is on the edge of its range. All species have a set of environmental conditions necessary for their survival. Climate is particularly important. For example, tropical plants and animals can only survive within a certain temperature range. As you move away from the tropics, they will gradually become rarer and then disappear. Even a slight amelioration of the climate such as we have seen recently in the northern hemisphere will enable them to expand their range. More temperate species will be adversely affected by rising temperatures, as well as increased competition from the invading southern species, so they may become scarcer. Another important climatic factor is water, both ground and atmospheric. Species adapted to very dry conditions will not survive in moist ones and vice versa.

Geology and soils are critical factors, especially for the nutrients and substrate they provide for plants, but, as animal species are ultimately tied to plants for their food source, these factors are vital for them, too. The underlying rock and the soil conditions, especially the degree of acidity, will determine which plant species are able to thrive in a given area. For animals, the soil or rock formations will often be very important in providing suitable refuges and breeding sites. Even fish will be affected by the surrounding rocks which control the acidity of and minerals in the water, as well as conditions at the bottom — a habitat important to many fish, especially when breeding.

These factors all interact with one another and, in some cases, the combination of conditions required by a species is very rare, so it will inevitably be confined to a very restricted area. Sometimes, rare species are relics left over from past ages which have been able to survive in isolated microhabitats. Still others may have survived from former climates. For such species, even very slight changes in local conditions will cause them to become first **vulnerable,** then **threatened,** then **endangered** and then **extirpated** or even **extinct.**

This is where human beings become a very important factor. Accidentally or on purpose, we are adept at changing environmental conditions. We drain swamps and irrigate deserts. We fell forests and plough up natural grasslands. We dam rivers and streams and "draw-down" wetlands. We discharge the waste products of civilization into waterways. We build roads and spew out gasoline fumes, and spread salt during the winter to facilitate our addiction to speed. All these factors combine to change not only the substrate but also the microclimate. We introduce domesticated and exotic species. Throughout the history of mankind, some species have been favoured and some extinguished. We occupy a much greater amount of space than would be normal for a creature of our size. This process has accelerated over the last century, partly because of the exponential rise in human population and partly through the use of machinery to change environmental conditions faster than ever before. The species most complained about by farmers and gardeners, i.e. *weeds*, are the ones for which we have created perfect conditions. Human groups that move around the world *like* familiar species, both plant and animal, and introduce them to their new home, creating havoc among native species. They *hate* other species such as snakes and strive to eliminate them. As a consequence, since man appeared on the scene species have become extinct at a much faster rate, especially in the past two centuries. Various calculations have been made but a conservative estimate is that the present rate of extinction is 100 times greater than would be normal. This means that extinctions greatly outnumber the origin of new species.

In Bruce and Grey counties we have examples of all types of rarity. We have species which require very specific conditions and are only found in very confined localities. We have many species on the edge of their range. Some of them are much commoner further north and some further south, others are more common to the west or east. We have some species only found in very specific conditions around the shores of the Great Lakes. The present warming conditions are causing some new species to move into the area. One example is the Great Egret which now nests here. It is possible that northern birds like the Common Loon are adversely affected by these changing conditions. In the latter case, however, human activity is also involved in its disappearance from some of its previous haunts. It is easy to observe this phenomenon with birds, but it also happens, although more slowly, with plants, where it is less easily observed.

Nature is never static, but natural changes are generally slow. The short human life span often makes such changes hard for us to observe, but in some instances, especially where habitats have been changed, species can be seen to have disappeared from particular sites in quite a short space of time.

In this book, we hope to draw attention to the fact that in Grey and Bruce Counties we have an unusually high variety of natural habitats. The most dramatic feature is the Niagara Escarpment, which traverses both counties, winding from Collingwood to Owen Sound and Wiarton, passing Lions Head and extending on to Tobermory. The

Peninsula and the continuation of the escarpment in the islands of Lake Huron provide a link between the north and the south of this province. The rocky shoreline and cliffs created on the Georgian Bay side of the Bruce Peninsula contrast with the sandy emergent shoreline and dunes of the main part of Lake Huron. The area to the west of the escarpment is basically a dolostone plain with lakes, crevices, potholes, alvars, swamps and marl fens. In Grey and southern Bruce counties, as well as along the escarpment, there are other exposed dolostone or limestone areas, such as the Rocky Saugeen valley. Ancient coral reefs in southeast Bruce have produced low cliffs along the upper Teeswater River and interesting rock outcrops in the region of Formosa. Some important natural areas are the Greenock Swamp (the largest in southern Ontario), Bognor Marsh and the Huron Fringe sand plain (south of Port Elgin). Glacial moraines form uplands such as "The Klondike." Other glacial features are several drumlin fields, kettle lakes, spillways, till plains and eskers. Another characteristic of this area is the high percentage of forest cover which remains everywhere except in much of southern Bruce County.

Native species are particularly vulnerable to violent changes caused by housing development, road building, poor forest management practices, draining wetlands, disruptive recreational activities, and also to some agricultural practices. Fragmentation of natural areas and changes to the substrate can cause extirpation of many species. The idea that disappearing wild species do not matter, so long as short term economic goals are achieved, is a very dangerous one. We are gradually becoming aware that rapid extinction of species brought about by our activities may, eventually, lead to our own extinction.

Definition of Terms

Endangered species are protected under the Endangered Species Act, and varying levels of protection are provided at the provincial level for rare, vulnerable and threatened species under the Planning Act. The process of determining a species provincially rare, vulnerable, threatened or endangered is not an easy one. Detailed status reports on each species must first be written and approved before its status is assigned by the Committee on the Status of Endangered Wildlife in Canada (COSEWIC) and/or the Committee on the Status of Species at Risk in Ontario (COSSARO). This is a lengthy and expensive process so that many rare species currently do not have official status. Using a totally different system NHIC assigns provincial rarity (S or Subnational rank) status based on the best available scientific information. They evaluate provincial ranks on a continuing basis and frequently produce updated lists. Information can be obtained from: <http://www.mnr.gov.on.ca/MNR/nhic/queries/countquery.cfm> The status of species rarity is extremely important in the land use planning process in Ontario. Often studies that show regard for rare species must be done prior to development projects proceeding.

Important Note: Those undertaking these rankings depend on information from the public about sightings of these species. Anyone who believes they have located these or any other rare species should complete a copy of the form at the back of the book and forward it to their local MNR office or the NHIC. Local field naturalist clubs will also be interested and may be able to help with identification. Records of any species from township units not marked on the distribution map for a species would be welcomed by the Bruce-Grey Plant Committee and The Owen Sound Field Naturalists. This is especially valuable for nesting birds and would be passed on to the Royal Ontario Museum. Photographs are useful and precise locations vital. It is absolutely essential that no species, plant or animal, should be disturbed or interfered with in any way while making these records.

Four main ranking schemes are applied to wildlife in Ontario.
These are:

Global Ranks — GRANK

Assigned by the Association for Biodiversity Information (**ABI**) in Arlington, Virginia, these refer to the status of the species "within the total range of the taxon." In the official listings the symbol "T" (for taxon) is used when the plant or animal has subspecies or varieties. It is not unusual for one subspecies to be much rarer than the other(s). Several of our list fall into these categories, but as we are only considering that section of the total species we have converted the ratings to "G" in order to simplify matters. For those who need to be more precise, the exact ratings can be checked on Nature Serve http://www.natureserve.org

G1 Extremely Rare — 5 or fewer occurrences or very few remaining individuals, or especially vulnerable to extinction.

G2 Very Rare — 6-20 occurrences, or with many individuals in fewer occurrences, or vulnerable to extinction.

G3 Rare — 21-100 occurrences, or with large numbers of individuals in fewer occurrences. May be susceptible to large scale disturbance.

G4 Uncommon to Common — Usually more than 100 occurrences. Usually not susceptible to immediate threats.

G5 Very Common — Secure under present conditions.

GU Status Uncertain — The species may be of concern but due to low research effort or its cryptic nature the exact status has not been established.

COSEWIC Status — assigned by the National Committee on the Status of Endangered Wildlife in Canada.

EXP Extirpated — Any indigenous species of fauna or flora no longer known to exist in the wild in Canada, but occurring elsewhere.

END Endangered — Any indigenous species of fauna or flora that is threatened with imminent extinction or extirpation throughout all or a considerable portion of its Canadian range.

THR Threatened — Any indigenous species of fauna or flora likely to become endangered in Canada if the factors affecting its vulnerability do not become reversed.

SC Special Concern — Any indigenous species of fauna or flora that is particularly at risk because of low or declining numbers, occurrence at the fringe of its range or in restricted areas, or for some other reason, but is not a threatened species at present.

NAR Not At Risk — Species which have been evaluated but have not fallen into any other category. It should be noted that COSEWIC only evaluates rare species so that any species considered are probably under some degree of threat (personal communication — Mike Oldham).

Provincial Ranks — SRANK — assigned to Ontario native species by the Natural Heritage Information Centre (NHIC). These ranks are not legal designations, but are designed to set protection priorities for rare species and natural communities.

S1 Extremely Rare in Ontario — 5 or fewer occurrences in the province, or very few remaining individuals. Often especially vulnerable to extirpation.

S2 Very Rare in Ontario — 6 to 20 occurrences in the province, or with many individuals in fewer occurrences. Often susceptible to extirpation.

S3 Rare in Ontario — Between 21 and 100 occurrences in the province, or with numbers of individuals in fewer occurrences.

Note: S1-S3 are referred to collectively as "Provincially Rare."

S4 Uncommon to Locally Common — Apparently secure in Ontario, usually with 101 to 1000 occurrences in the province.

S5 Very Common — Demonstrably secure in Ontario, usually with greater than 1000 occurrences.

SX Extirpated — Apparently gone from the province with very little likelihood of rediscovery.

MNR Status — assigned by the Ontario Ministry of Natural Resources. Since January 1995 these are based on the recommendations of a Ministry technical committee called the Committee on the Status of Species at Risk in Ontario. (COSSARO)

END Endangered — Any indigenous species of fauna or flora threatened with immediate extirpation throughout all or a significant portion of its range in Ontario.

These species are protected under the Endangered Species Act of the province of Ontario.

THR Threatened — Any indigenous species of fauna or flora indicated to be experiencing a definite non-cyclical decline throughout all or a major portion of its Ontario range, and that is likely to become endangered if the factors responsible do not change.

VUL Vulnerable — Any indigenous species of fauna or flora represented in Ontario by small but relatively stable populations and/or which occurs sporadically, or in a very restricted area, or at the fringe of its range. Periodic monitoring for possible decline is necessary.

Note: In general, only species with a classification of S1 to S3 that have been seen in Grey-Bruce recently are fully described. Other species classified as rare by NHIC which are historical, possibly not native to the area, or for some reason doubtful are listed in separate sections. There are many other interesting species which are rare in this area but space does not permit us to describe them here. It is probable too that some of them, especially fauna, should be classified, at least provincially, but not enough is known about them.

Bruce and Grey Counties — Past and Present

445 to 420 million years ago — from the Upper Ordovician to the Middle Silurian Periods

The land is silent, except for the wind and the sound of rain rushing off the bare Taconic Mountains to the east, flowing into the gurgling streams and moving on to the sluggish waters of the river deltas where the burden of sand, silt and clay drops among the seaweeds and teeming populations of slithering worms and jointed armour-plated invertebrates. Year by year the layers of sediment build up imprisoning the corpses of the inhabitants of this warm salt water, whose remains may be seen today in the rock slabs on the shores of Georgian Bay. Corals start to form a reef around the edge of the sea in the Michigan Basin, where Lake Huron and Lake Michigan are now found.

420 to 245 million years ago — Late Silurian to Early Permian Periods

Sediments continue to accumulate in the Palaeozoic seas, layer upon layer of shales, silts, gypsum and carbonates, burying the rocks that are exposed today under a great thickness of sediment, the earth's crust sinking beneath the weight. Finally, the Taconic Mountains have been laid low and the Appalachians begin to build in their place.

245 million to 1 800 000 years ago —
Permian through Mesozoic Era to the Pleistocene period of the Cenozoic Era

The land rises and the sea dries up, although the salt is still there beneath more recent deposits as well as the oil formed from the remains of the plants and animals of long ago. Immediately the forces of erosion start to work on the newly emerged land. The soft sediments are worn away during millenium after millenium of erosion. The first land plants that appeared in the Silurian have proliferated into innumerable species. The giant horsetails and ferns of the Paleozoic are displaced by the flowering plants. Forests of trees as we know them appear and new forms of animal roam among them.

1 800 000 to 10000 years ago — The Pleistocene Ice Ages

The climate becomes colder, ice sheets start to form, gradually moving south, until a layer two to three kilometres thick covers the two counties. The ice comes and goes many times. The final retreat leaving the landscape we know today, with rocks — erratics — moved from far away and vast deposits of sand and gravel. The ice and meltwater create the drumlins and moraines. Other ridges — eskers — form by deposition from subglacial rivers. Water pouring off the melting ice cuts channels in the previous deposits and spreads new layers of sand and gravel over the land. Water levels in the Great Lakes fluctuate wildly as meltwater fills them and different outlets open and close.

From beneath the ice and water emerges a cliff showing the old sedimentary beds, forming what we call the Niagara Escarpment. The waves of the postglacial lakes undercut caves on its cliffs but then, as the lakes recede, the caves are stranded high above modern lake levels, along with boulder beaches, and the gravels and clays of the old lake beds which slope gently to the shores of modern Lake Huron. Wetlands, sometimes with dry hummocks that are drumlins or other glacial landforms; the meandering rivers, often in old glacial channels; and abandoned beaches and beach bluffs are all that remain of the vast post-glacial lakes. The high calcium content of the Paleozoic rocks affects the open wetlands so that many are fens; true acid bogs are rare.

10000 to 160 years ago

First the lichens and mosses invade the newly exposed rocks. Other plants move into the wet hollows left as glacial Lake Algonquin retreats, gradually filling them with peat as the centuries pass. Then the willows and dwarf birches appear, followed by the trees that we know today with all the plants and animals associated with them. Thus a new landscape is created. Fish of all sizes and shapes swim in the streams and lakes. Humans, too, enter the scene. At first, they hunt the large mammals on the edge of the ice and trap the smaller ones in the mossy hollows. Then, as the forest grows, they adapt to new forms of food. They hunt elk and deer, catch passenger pigeons that roost in huge numbers in the massive trees, gather herbs in the spring, harvest berries in the fall, and use the products of the forest to build their dwellings. They start to cultivate small patches of corn, beans and squash, and develop a more settled life style in small villages. Every 15 to 20 years the villages move to another area with fresh soil and an accessible supply of firewood. The Missisauga-Ojibwa people have long established themselves in the area by the time the first French explorers and priests appear. They live a peaceful, well organized life, using waterways to travel in summer and snowshoes in winter. They have learned how to survive long hard winters and how to live in harmony with the natural world. The land does not belong to them, they belong to the land. The area is full of variety, from the escarpment with its cliffs and caves and fissured surfaces, to dry upland forest grading into moist lowland forest, swampy forest and lowland fens. Shoreline dune-panne complexes are found on the Bruce Peninsula as well as flat, open limestone pavements. The deciduous forests are open with enormous widely spaced trees, trunks up to 2 metres wide and towering 30 metres into the air. Cedar, elm, white pine, and tamarack grow in the wetter areas; sugar maple, beech and hemlock where it is better drained. The forest floor is cluttered with deadfall and rotting logs, smaller trees cluster below the giants waiting for nature to fell the ancient one after its 300 year of domination. Along the waterways, giant cedars arch over, and on the escarpment, small gnarled ones grow for over 1000 years.

Gradually, rumours begin to circulate among the indigenous people of strangers with pale faces and black robes living among their neighbours across the bay, but more pressing problems concern them. The aggressive Iroquois are pushing up from the southeast into the traditional lands of the Ojibwa in the middle of the seventeenth century, disrupting the trade routes that have been established since the coming of the French and destroying the villages. Those that escape are forced to retreat to

Manitoulin Island in the north. No longer can the northern Ojibwa portage from Lake Huron through Boat Lake to Colpoys Bay to avoid the dangerous waters around the north of the peninsula. The Iroquois rule the whole area for the next 50 years. In 1690 the Ojibwa make a return. A decisive battle is fought just outside modern-day Southampton. Even the women and children help decoy the Iroquois into the trap; so many are killed on both sides that the bodies are left to decay and the historic event becomes known as the Battle of Skull Mounds. The Ojibwa re-establish small communities in the southern parts of the two counties and their hunting grounds on the peninsula. There they live in relative peace for over 100 years more. In 1768 the total population of Missisauga-Ojibwa from Lake Ontario and Lake Erie to Lake Huron including Georgian Bay is estimated to have been 2700.

1840 to 1940

For hundreds of years the shoreline of the Bruce Peninsula was known as a bleak and rugged and dangerous coast, but to Europeans it is an empty land ripe for settlement. In 1840 George Rankin arrives to survey the area around what is now Owen Sound, where the Nawash village is home to about 180 people. He divides the map into neat rectangles of 100 acres each, totally ignoring the topography and the ecological integrity of the area. A similar process soon takes place in the "Queen's Bush" where the Saugeen Indians live. The Garafraxa Trail marks a straight line through forest and wetland linking the older settled areas to the south with this new conquest to the north. The Durham Line branches off to link up the Queen's Bush. Before the summer of 1840 is up, land agent John Telfer has sailed into Owen Sound and built himself a house, so that he can distribute the land fairly to the new settlers who start to arrive with their bundles of worldly possessions, their unsuitable clothes and their lack of experience of this country. These were people who had always made an enemy of the forest and of the wolves and bears found there, but there was land to be had, and trees to be cut for profit, animals to be trapped for fur, and fish to be caught, crops to be planted, houses to be built, industry to develop, and money to be made. With the necessity for personal survival this was all done with an almost total disregard of the elemental forces of nature that had created this land and its unique living things, and the people who had lived here for ten millennia. This was an opportunity to get out of overcrowded Europe and to do what the European tribes had always done when they needed more room — move in and take over. The indigenous people were promised lands. First of all, they were given the whole of the Bruce Peninsula, but throughout the nineteenth century these grants were gradually reduced in area, until they were marginalized on the Saugeen and Cape Croker Reserves with only small areas reserved for their hunting grounds. As Sir Francis Bond Head summed it up in an 1836 treaty " In all parts of the world farmers seek for cultivated land as eagerly as you, my red children, hunt in your forest for game. If you would cultivate your land it would be considered your own . . . but uncultivated land is like wild animals, and your Great Father, who has hitherto protected you, has now great difficulty in securing it for you from the whites who are hunting to cultivate it."

The first farms were in tiny openings in the forest. Their log houses had earth floors and were roofed with elm bark or split basswood. Within a few decades, large spreads of land were cleared and farms and orchards were thriving, towns were developing, loggers and industries such as furniture-making based on their products were creating fortunes, roads criss-crossed the landscape, trapping had removed the beaver and most fur-bearing species, some fish species were on the verge of extinction and new ones were being exploited, rivers had been tamed and swamps were being drained. The human population had risen from a few hundred to several thousand. By 1871 the population of Grey County alone was 59 395. From the 1920s summer cottages began to line the shores of the lakes, especially where the beaches were sandy on Lake Huron. The Bruce Peninsula was becoming a summer destination.

1940 to the Present

The Second World War brought dramatic changes. The world economy opened up and marginal farms were no longer profitable. The loggers had moved on long ago, and raging fires swept the Peninsula at the turn of the century. The ruggedness of the terrain and the prevalence of wetlands had prevented the land ever being totally cleared, although it was almost all logged at some time. As the farms were abandoned the forest began to creep back into the cleared areas, overtaking the former orchards and fields, but periodic logging continues. By a strange twist of fate a huge area near Meaford was cleared of farms so that the army could train for the war in Europe. Today, only traces of the old settlements remain and there has been much regeneration in the parts the armed forces do not use. In other places, where the soil was good, the farms still flourish, especially in southern Bruce County. Around Thornbury and Meaford a thriving apple growing industry has developed. Compared with many areas of Ontario, the two counties have been fortunate in that samples of all the different kinds of natural environments have survived, although fragmented and reduced in size. We are beginning to recognize that if you respect the environment and leave things alone nature will take care of itself.

References

Barry, James. 1968. **The Sixth Great Lake**. Clarke Irwin, Toronto.

Chapman, L.C. & D.F. Putnam. 1984. **The Physiography of Southern Ontario**. On. Geo. Sur., Ontario MNR.

Fox, W. Sherwood. 1952. **The Bruce Beckons.** University of Toronto Press.

Schmalz, Peter. **The Ojibway of Southern Ontario**.

Marsh, E.L. et al. 1931. **A History of the County of Grey.** Fleming Pub. Co. Owen Sound.

Miklos, Helen H. & T. Pinther. 1987. **Atlas of the Great Lakes Indians.** Univ. of Oklahoma.

Robertson, Norman. 1960 revised. **The History of the County of Bruce.** J.M. Dent, Canada.

Tovell, Walter, M. 1992. **Guide to the Geology of the Niagara Escarpment.** Niagara Escarpment Commission.

Natural Habitat Types in Bruce and Grey Counties

These habitat types do not occur in isolation; one merges into another. This outline is intended to give those unfamiliar with the area some idea of the great variety of habitat available to both plants and animals.

A. Forests:

Forests with more than 75% broad-leaved (deciduous) trees are called **deciduous forests.** Forests with more than 75% cone-bearing trees are called **coniferous (conifer) forests**. Forests with a more equal mix of deciduous and coniferous trees are called **mixed**.

a. Lowland Forests

These are natural woodlands on wet or moist ground (including areas termed as swamps, swamp forests, and swampy woods or forests).

1. **Lowland Deciduous Forest** (including deciduous swamps)
   The most abundant trees overall are the "soft maple complex" (Silver and Red Maple) and also Red (Green) Ash *Fraxinus pennsylvanica*. In a broad section of Grey County, this ash is not abundant. Sometimes Soft Maple is absent, especially in Bruce County.
2. **Lowland Mixed Forest** (including mixed swamps)
   White Cedar is the most abundant tree overall. Many others are common.
3. **Lowland Coniferous Forest** (including conifer swamps).
   Again, White Cedar is the most abundant tree overall. Tamarack is the second most common, but not in more acidic sites — termed by many as "wooded bogs."

b. Upland Forests

This includes all natural woodlands not moist enough to be termed lowlands — regardless of the actual elevation in relation to adjacent areas.

1. **Upland Deciduous Forest**
   Sugar Maple is the most abundant tree overall, although it is not dominant in some locations, and may even be scarce or absent. Other very common trees are Beech, White Ash, White Birch, Black Cherry and Basswood. This is the most abundant type of natural habitat in Grey and Bruce Counties.
2. **Upland Mixed Forest**
   White Cedar is generally the most abundant tree, with White Birch the second commonest. Many other tree species will be found.
3. **Upland Coniferous Forest**
   White Cedar again is generally the most abundant tree, but there are also areas dominated by Jack Pine. The latter areas are home to a number of provincially rare plant species.

B. Open (unwooded) Areas.

Natural open areas are much less common in this region than woodlands. Open areas are home to many of the provincially and nationally rare species. Trees may occur sparingly but they are widely spaced and often stunted in growth.

a. Wet Areas

1. Fen

An open, non-acidic wetland with herbaceous vegetation that is not tall. Commonly, there is very slowly flowing or seeping water down very low gradient slopes; sometimes the vegetation forms a shallow floating mat. In this area fens are normally calcareous, often marly. In general, members of the Sedge family dominate.

2. Marsh

An open wetland with tall herbaceous vegetation, and often anchored floating plants and submergents. Usually flooded seasonally, sometimes throughout the year. Coarse members of the Sedge family are generally most prominent in Grey and Bruce Counties, although there are exceptions.

3. Bog

An open acidic wetland where herbaceous vegetation is not tall (tree cover does not exceed 25%). Normally, there is little if any obvious inflow of water; most of the moisture comes from precipitation. Peat mosses (*Sphagnum*) and woody members of the Heath family, such as Leatherleaf, predominate.

Note: In this area Bogs are rare because of the calcareous bedrock which lies near the surface. Both Marshes and Bogs are much scarcer than in other parts of Ontario. The reverse is true of Fens which are much commoner here than elsewhere. Some wetland areas are intermediate to some extent, and experts classify them under other terms such as "poor fen" which are difficult to define and confusing to non-professionals.

b. Better Drained Areas

1. Meadow

A natural grassy opening. Most of the meadows here are moist, marly sites close to the shore of the main body of Lake Huron. They could be described as "less wet fens."

2. Alvar

A flat open area with calcareous bedrock at, or just below, the surface — normally at least partly exposed. Most of the alvars have dolostone pavements, but **prairie alvars** on shale or limestone also occur. A number of provincially rare species, including the Eastern Massasauga Rattlesnake, occur in alvars and the adjacent forests.

3. Open (Unwooded) Sand

This largely takes the form of open sand dunes (sand hills or undulations) along or near Lake Huron — only rarely on Georgian Bay. Grasses (often very tall) tend to dominate. Many provincially and nationally rare plants occur in these locations.

4. Open Shore

Much of this heterogeneous habitat is found along Lake Huron, including Georgian Bay. Exposed (untreed) islets, open sand dunes, fens, meadows and alvars, as well as various types of beach are included. A high number of rare species occur.

5. Cliffs and Open Talus Slopes

A high vertical exposure of ground and adjacent unforested rock fall. In Grey and Bruce Counties these areas are found along the Niagara Escarpment and are all dolostone. Mosses, lichens, ferns, stunted trees (sometimes ancient cedars) and other plants occur.

Note: Even with these definitions of rocky areas, there are some open bedrock locations which elude classification as cliff, alvar or shore!

C. Aquatic Habitats

A great variety of highly adapted plants and animals occur in these aquatic habitats. In addition they are essential to many land animals.

1. Lakes and Natural Ponds

There are almost 100 inland lakes in the two counties. In these and smaller ponds, White and Yellow Water Lilies and Hardstem Bulrush are common.

2. Lake Huron including Georgian Bay

Habitats range from shallow waters with a sandy bottom or rock pavement, to water more than 200 metres deep.

3. Rivers and Creeks

These abound throughout the area and range from calm, slow-moving stretches (often deep) to rapids and waterfalls (coming off the escarpment). Bottom conditions range from mud to sand or gravel or rock, and will often vary in close proximity providing breeding and feeding areas suitable for different species.

Man-Created Habitats

Note that a few rare species seen in this area occur in man-created habitats such as abandoned fields — often with shrubby vegetation, or fields currently being cropped or pastured. Some occur on roadsides, in conifer plantations and on lawns.

Abbreviations and Measurements Used in the Descriptions

At the top of each species description page, where they are applicable, the abbreviations for the Global rank, the COSEWIC status, the NHIC "S" rank and the MNR status are listed in that order. Not all species are ranked in all four categories; frequently there is no MNR status.

Metric measurements have been used throughout this book. For those unfamiliar with this system, the following approximations may be helpful:

10 mm = 1 cm; 25mm = 1 inch
2.5 cm = 1 inch; 10 cm = 4 inches; 30 cm = 1 foot
1 metre = 39 inches i.e 3 inches more than 1 yard
1 km = 5/8 mile or 1.6 km = 1 mile

Map of Township Units — Based on Historical Townships

Note:
The square is marked if the species has been found at least once in that township unit. Every endeavour has been made to ensure the authenticity of these records.

If anyone discovers in a township unit a species that is not recorded here, please inform the M.N.R. (See form at the back).

For explanation of Township Units, see Appendix I.

FISH

Grey County, and especially Bruce County, have extensive Great Lakes coastlines, and the surrounding waters are home to a variety of fish. The Lake Huron shore in Bruce County tends to be sandy with shallow water, or with dolostone shoals. In Georgian Bay the shore is more often rocky and there is often a steep drop-off into very deep water. There are many rivers and streams running into the lakes; some, such as the Saugeen, large and meandering; others near the escarpment fast-running with falls, rapids and gorges. Conditions at the bottom of the rivers, which are often important for spawning fish, range from muddy to sandy to gravelly to rocky. These varied environments support a variety of aquatic species. As with all ecosystems, different species occupy different niches and rely on different food sources. Similarly, there is a "food chain" with small species in large numbers dependent on plants and invertebrates, which in turn are preyed upon by larger carnivorous species. The aquatic ecosystem is not independent of the land. The surrounding geological features affect the quality of the water. The level of nutrients, both organic and inorganic, is affected by run-off from the land.

The people of North America have used these waters as a food source for millennia, and when the Jesuits arrived in the mid-1600s they found Lake Huron, Georgian Bay and the rivers teeming with tasty fish. As European settlements developed, especially in the 19th century, it was not long before commercial exploitation of the fish stocks began in Lake Huron and Georgian Bay. Many other changes started to take place. The growth of towns; the development of agriculture and industry; the installation of dams; the use of coal powered, and later, petroleum powered vessels; all helped to change the nutrient status of the water and the fish habitat. In this area the impact has not been so great as near larger cities. We still have a relatively small population in small communities, and little industry, although agriculture is an important feature of this region. With the Europeans also came the introduction of exotic species, sometimes accidentally, sometimes on purpose. More recently, excessive recreational use and cottage development along shorelines can have a damaging effect.

Under all these changing circumstances, the rarer native species tend to suffer most because they are the ones with the most specific habitat requirements, and are least likely to survive extensive pollution and/or overfishing. Even minor changes can drive them to extinction. Over-exploitation of highly prized species can also lead to extinction if there are not adequate controls. The Lake Sturgeon is an example of such a species which was decimated not only by overfishing, but also by habitat loss, the construction of dams and pollution. It has, hopefully been pulled back from the edge of oblivion by suitable conservation measures. Eight rare fish species which inhabit these waters are listed here. Most of us are never likely to see them but their presence or future absence is an indicator of the health of our environment as well as theirs. Mention is also made of a number of species once favoured as a food source which are no longer here.

Note: Among the species described in this section is one lamprey. Lampreys are often included with fish because they share a similar environment. They belong to a very primitive group of vertebrates — the Class Agnatha or Jawless Vertebrates. They have no proper spine; the primitive notochord persists in the adult. The skeleton is cartilaginous. Unlike bony fish, they have no gill covers. The gills are visible as a row of slits on each side of the eel-like body. There are no paired fins, only one fin running from halfway along the back around the rear of the body. They have no scales, the body is coated with slime. They have a life cycle like many invertebrates with a larval stage and metamorphosis into the adult.

References

Ontario's Biodiversity Fast Facts. Royal Ontario Museum & MNR. http://www.rom.on.ca/cgi-bin/cbcb/fastfact

Campbell, Neil A. 1993. **Biology**. 3rd.Ed. Benjamin/Cummings Publishing Co. Inc., Don Mills.

Scott, W.B. & E.J.Crossman. 1998. **Freshwater Fishes of Canada** Dept. of Oceans & Fisheries. Bull. 184.

1. Northern Brook Lamprey G4 SC S3 VUL

***Icthyomyson fossor* Reighard & Cummins**

A great deal is heard about the Sea Lamprey, which gradually moved into the upper Great Lakes by way of the Welland Canal, although it took 100 years before it became a serious problem. This parasitic species has had a serious negative effect on indigenous fish species, but it is comparatively rare in this area now. It is not generally known that there are several other species of native lamprey which are not necessarily parasitic. The Northern Brook Lamprey is such an example. Typically, the lampreys look like a small eel with a round mouth and teeth arranged in a circle. This enables them to fasten on to their food source and extract the juices. The Northern Brook Lamprey has these characteristics, but curiously the adult never feeds, even though it lives for about a year before spawning and dying. One female may lay as many as 1000 eggs in a "nest" the pair create in the gravelly stream riffles they prefer. When the larvae hatch out they burrow into the soft mud at the bottom and live for about six years, feeding on microscopic diatoms, protozoa, algae and dead organic detritus. They are worm-like, blind and toothless, and so different from their parents that, for a long time, they were thought to be a different species. At maturity they change into an immature adult but overwinter in the mud for another eight months, before finally emerging and dispersing to their spawning grounds. The typical habitat is small rivers, and there are obviously very specific habitat requirements, so that factors creating environmental changes make them vulnerable. Siltation and pollution causing changes to the bottom and food sources affect the larvae. However, the greatest danger comes from efforts to control the Sea Lamprey which also breeds in small rivers around the Great Lakes. Application of non-selective chemicals (lampricides) to these streams kills native Lampreys as well as the target species. Fortunately, only the lower reaches are treated so the Brook Lamprey should escape as it usually lives further upstream. In this area it can be found in both Bruce and Grey Counties. While some may ask why we should be concerned about such a seemingly insignificant creature of this nature, it should be remembered that they are near the bottom of the food chain. They process minute organisms and provide food for larger creatures. Removing any species reduces biodiversity and creates a gap in the ecosystem which can affect many other creatures.

DESCRIPTION

Length and Shape: 6-16 cm. Cylindrical, eel-like.

Colour: Brownish-grey.

Head: Terminating in a circular sucker lined with blunt, horny teeth.

Fin: One, dorsal, narrow with one undulation, terminating on lower side at back. No fin-rays.

Skeleton: Cartilaginous.

Spawning: June — water temperature ± 14°C

Habitat: Small rivers and creeks. Sometimes in large rivers or lakes or small pools.

Range: Eastern United States and Canada. Upper Mississippi and southern Hudson Bay watersheds, and Lakes Superior, Huron and Erie, and the Ottawa and St. Lawrence Rivers.

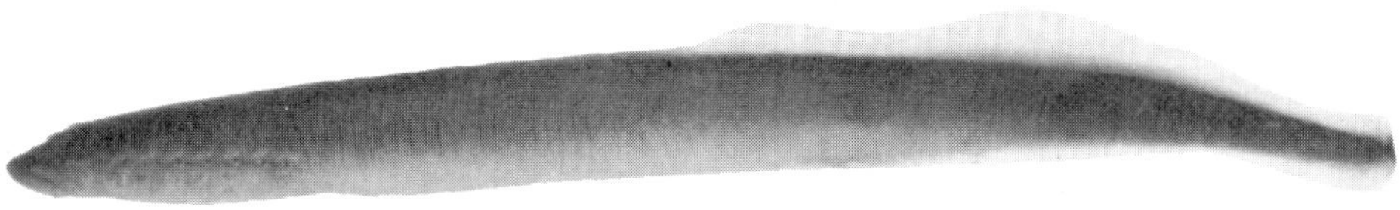

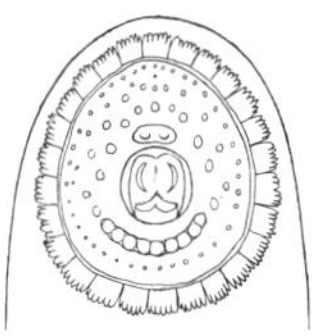

Class: Osteicthyes — Bony Fish
2. The Ciscos

Shortnose Cisco	***Coregonus reighardi*** **Koelz**	**G1 THR SX EXP**
Shortjaw Cisco	***Coregonus zenithicus*** **Jordan & Evermann**	**G2 THR S2 THR**
Blackfin Cisco	***Coregonus nigripinnis*** **Gill**	**GX? THR SX EXT**
Kiyi	***Coregonus kiyi*** **Koelz**	**G3 SC S3 THR**

The eight species of Cisco found in the Great Lakes are herring-like, medium sized fish, found in relatively deep water. They are often referred to as Chubs. They have all suffered from commercial overfishing, having been regularly harvested in the 19th and early 20th centuries until about 1950. The introduction of the parasitic Sea Lamprey, and other invasive species which compete for limited available food sources, such as the Alewife and Rainbow Smelt, was also a factor in the decline which accelerated in the middle of the last century. The four species of Cisco listed here are considered to be threatened in Canada, and in three cases are already extirpated in this area. The Shortnose Cisco was the only one believed to have a remnant population in Georgian Bay, but recent efforts to capture them have failed and the MNR have just listed them as extirpated. There is still no specific legal protection from commercial harvest. The Shortjaw, length about 27 cm, was last seen in Lake Huron in the early 1980's. In Ontario today it is only found in Lakes Superior and Nipigon (a different form from Lake Huron) and some smaller inland lakes. It also occurs in the western provinces. The Blackfin was last seen in Lake Huron in 1960. It has disappeared in Michigan. The only known remaining population is in Lake Nipigon. This silvery herring-like fish, with its black fins and large eyes, lives at depths of up to 180 metres. The Kiyi is rather smaller than the others, only attaining a length of about 25 cm. Nevertheless, it was fished so intensively that it has not been seen in Lake Huron since 1973 but it still exists in Lake Superior. Pollution and eutrophication enhanced by the discharge of sewage, as well as agricultural and industrial waste, are most likely other factors in the decline of these species. The decay of the resultant additional surface vegetation such as algae depletes oxygen supplies at the lower levels which these fish inhabit. It should be noted that these moderately sized fish provided food for larger predators, so that their decline has also affected other species.

Shortjaw Cisco

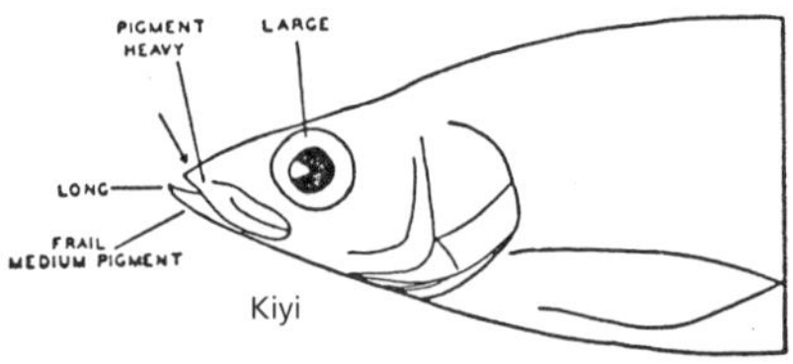

Kiyi

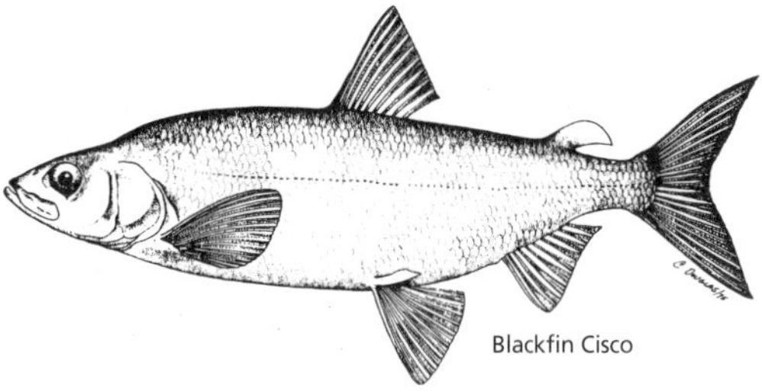

Blackfin Cisco

Shortnose Cisco

DESCRIPTION
Length and shape: 25 cm. Widest in front half. Tapering to forked tail. Only lightly compressed laterally. Head short and stout.
Colour: Silvery, iridescent sides. White below.
Fins: Dorsal — small, central, very small dorsal fin in front of tail. Pectoral pair very small and low. Pelvic pair very small. Anal fin small. Tail large and forked.
Spawning: Spring — possibly also Fall. Over sand, silt or clay bottoms.
Food: Copepods, aquatic insect larvae, fingernail clams.
Habitat: Depth 23-90 metres.

3. Lake Sturgeon G3 NAR S3

Acipenser fulvescens **Rafinesque**

The Lake Sturgeon is a large fish which is endemic to the large rivers and lakes of eastern and central North America as far west as the North Saskatchewn River in Alberta and as far north as Hudson Bay. In the Great Lakes, it was said that Lake Sturgeon approaching 300 lbs (140 kg) and 7 ft (2 m) in length were formerly caught. Such large fish are now extremely rare. It is a very strange looking creature as, instead of scales, it has tiny toothlike notches over its very tough skin. In addition, there are rows of bony plates along the body. Altogether it is a well armoured fish, and one that had a bad reputation among early settlers for destroying fishing nets. It was also considered to prey on the eggs of other commercial fish. In fact, it is a bottom feeder living largely on invertebrates, and the consumption of eggs is more or less accidental. It took some time before it was realized that the fish they discarded (either burning, dumping, or feeding to the pigs) was one of the most commercially valuable. The roe was used for caviar, the flesh eaten smoked or fresh, isinglass (used as a clarifying agent, as cement for pottery, as a gelling agent, or for waterproofing) was extracted from the swim bladder, and the skin was tanned for leather. By 1885 Lake Sturgeon was one of the most valuable products of the Great Lakes fishery. The intense exploitation of this slow growing fish resulted in a serious decline within ten years and this continued throughout the twentieth century. Another factor affecting this species is the damming, and degradation through deforestation, agriculture and pollution of the tributaries where they spawn — often far up-river — and where the young may live for up to a year. In recent years, there has been much interest in the survival of this unique fish. Oddly enough, the spread of the Zebra Mussel may actually be of benefit to it as a new food source. Current efforts to control pollution and improve fish habitat seem to be paying off, and there are signs that the population is increasing.

Local Status: Lake Sturgeon seem to be increasing. Commercial fishermen based in Collingwood are reporting increased catches of Sturgeon of various age classes. This is a good sign as it means they are actively reproducing. However, they typically prefer deep, fast flowing rivers for spawning and these are uncommon in Grey and Bruce Counties. There is a report from the Sauble River in 1999.

DESCRIPTION

Length and Shape: ± 110 cm when fully grown. Tapering oval body with a long snout.
Colour: Olive-brown to grey, white below. Fins dark brown to grey. Black blotches on young.
Fins etc.: Dorsal fin near tail. Pectoral pair medium sized. Pelvic pair small. Larger anal fin close to tail. Tail unequally forked. Spines along back. Horny plates along sides, back and underside.
Spawning: June. On clean, gravel shoals and in stream rapids. Females mature between 14 and 33 years, spawning every 4 to 9 years. Males mature 8-12 years, spawning every 2 to 7 years. Only 10-20% spawn in a given year.
Life Span: Females 80-150 years. Males 55 years.
Food: Insect larvae, crayfish, snails, clams, leeches.
Habitat: Near lake shores, water 5-10 metres deep.

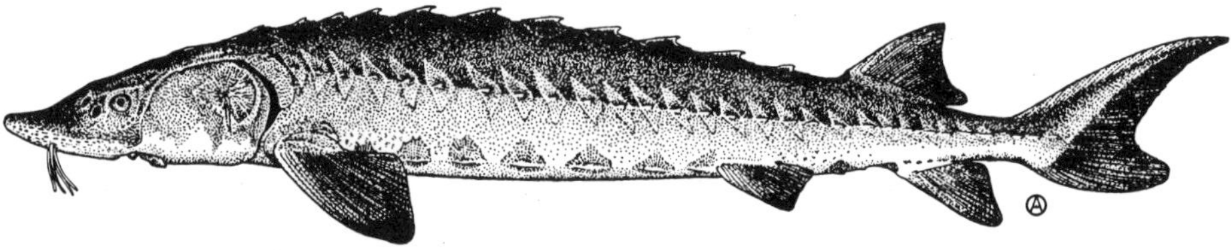

4. Redside Dace **G4 SC S3 THR**

Clinostomus elongatus

The Redside Dace is a small fish which inhabits streams in eastern North America. It is a member of the Minnow or Carp family, to which the Asian Goldfish also belongs. This colourful little fish is iridescent blue and green, with a golden stripe along the sides and below that a red band stretching from the gill cover to below the dorsal fin. In Scott and Crossman (1973) it was believed to be found only in streams flowing into western Lake Ontario. Unfortunately, in the past, fish research often seemed to centre on the commercial or sports fisheries — fish that "were useful to man." A small fish like this, which had no "value" from that point of view, often escaped attention. However, now that we are beginning to take a more ecological viewpoint, it has been recorded from more than 200 sites in Lake Erie and Lake Huron streams as well as Lake Ontario. Redside Dace were found in the 1950s at a number of sites in the headwaters of the Saugeen River and on Meux Creek, where more recent surveys have failed to locate them. It is estimated that there has been a 30% decline in its range in this area. There seems to be some hope that they might be recovering as Erling Holm captured 6 specimens in the upper Saugeen River in 1999 at sites where they had been present in 1972 but not in 1985 (personal communication). Streams with clear, moving water and a gravel or stony bottom are preferred. They often seem to spawn in the breeding pits made by Creek Chub, even when the male chub is present (Scott & Crossman, 1979). They are vulnerable to stream siltation due to erosion and the removal of streamside vegetation. They are insect feeders and may often be seen leaping out of the water to catch flying insects. Pesticides used in agricultural operations alongside the streams, which destroy the insects on which this species feeds, also pose a threat. Restoring streams to a more natural state with flourishing streamside vegetation is the best way to assist the survival of this interesting small creature.

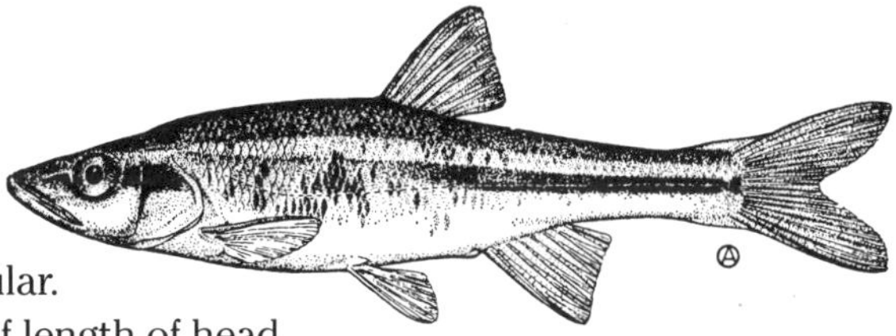

DESCRIPTION

Length and Shape: ± 8 cm long. Slender, elongate. Laterally compressed. Head pointed, triangular. Large eye. Large mouth, almost half length of head.

Colour: Back dark-green/blue-green, iridescent. Sides with yellow stripe and red band, somewhat speckled. Lighter below. Scales smooth ± elongate.

Fins: Dorsal — large, central. Pectoral and pelvic pairs — small. Anal — large, one third of body length in front of tail. Tail — large, symmetrically forked.

Spawning: June. In riffles, shallow flowing pools and Creek Chub nests.

Life Span: 4 years, breeding in the 3rd year.

Food: Mayflies, dragonflies, spiders.

Habitat: Cool, clear, flowing water with gravel or stony bottom.

5. Deepwater Sculpin **G5 THR S4**

Myoxocephalus thompsonii **Linnaeus**

The Deepwater Sculpin is a tiny, freshwater species related to the marine Fourhorn Sculpin, and distinguished from the latter by living in deeper water — hence its English name. This fish is a circumpolar species, which is found in freshwater lakes in an arc across Canada from the Great Lakes to the Northwest Territories. It is believed to be a post-glacial relict which was pushed south during glaciation, separated from the marine species, and moving north again as the ice sheets retreated. At one time it was considered to be a nuisance by fishermen in Lake Ontario as it often became entangled in gillnets set for Lake Trout. Pollution in the sediments in Lake Ontario and Lake Erie, causing reduction in oxygen levels and food sources, produced a dramatic decline and it was believed to have been extirpated by the 1970's. However, it has recently been rediscovered in Lake Ontario, an indication that there has been an improvement in the quality of the lake and it may reappear in Lake Erie. Its numbers seriously declined in Lake Huron, although it did not disappear. However, according to the Lake Huron Management Unit of the Ministry of Natural Resources it appears to be increasing in numbers again. Deepwater Sculpin are very strange-looking fish with large heads and fins. The average depth of water which they inhabit is about 70 to 100 metres, but in Lake Superior they have been taken by trawl nets at 400 metres. They feed on invertebrates found in the mud at the bottom. Their habitat is difficult to access, except by skilled divers, so they have not been well studied. There is some evidence that there is parental care of the eggs. They are considered to be a valuable food source for species such as Lake Trout which are prized by fishermen, so that, once again, removal of an "insignificant" species may result in the decline of a food source for us.

DESCRIPTION

Length and Shape: ± 10 cm. Elongate, heaviest in the front half, tapering to the rear. Head large ± 28% of body length. Eyes on top of head. Snout rounded, mouth large.

Colour: Grey-brown. Speckled. 4-7 saddle-like bands in rear half.

Fins: Dorsal — 2, forward one half size of rear one. Pectoral pair — large, fan-like. Pelvic pair — small. Anal — large. Tail — medium size, fan-shaped.

Spawning: Summer or early fall.

Life Span: Not well understood.

Food: Insect larvae and other invertebrates.

Habitat: Bottom of deep (50 m and below), cold lakes.

6. Longear Sunfish G5 NAR S3

Lepomis megalotis **Rafinesque**

This small, colourful, almost circular fish really is reminiscent of the sun and its rays, with the addition of a distinct "ear" at the back of the gill cover. The thirty species of Sunfish are exclusively North American, mostly found in the Eastern half, extending into Mexico. Nine species are found in Canada and these include three species of Bass (Rock, Smallmouth and Largemouth) and the Pumpkinseed. The Longear Sunfish occurs in parts of the Upper St. Lawrence River in Quebec. In Ontario, it has been found in Lake Erie, Lake St. Clair and the southern Georgian Bay area, with a disjunct population in Rainy River District. It is an inhabitant of shallow water in medium sized rivers, ponds and small lakes. It has never been very common and has not been well studied. The adults come to maturity at 2-4 years of age and live potentially to 8 years. The males exhibit elaborate territorial behaviour. They build a shallow, saucer-shaped nest in gravel, or failing that in sand or mud. After the female lays the eggs and they are fertilized, the male guards the nest and fans the eggs for the 3-5 days before they hatch. He stays on guard over the tiny fry for a few more days before leaving them to their own devices. They seem to feed largely on mature insects, so spend much of their time near the surface of the water. They also eat other invertebrates and, sometimes, small fish.

Local Status: Reported from the Bruce Peninsula (Erling Holm ROM).

DESCRIPTION

Length and Shape: ± 11 cm. Almost as wide as long. Strongly laterally compressed. Head triangular. Eye large. Mouth fairly large, slightly up-tilted.

Colour: Olive to rusty brown, spotted with orange and emerald to turquoise. Cheeks orange with wavy blue streaks. Breast and lower surface pale red, or orange to yellow. Fins brownish to orange, black at base. Ear flap dark brown or black, narrowly edged with pale red to yellow.

Fins: Dorsal — 2 joined, front spiny, back larger. Pectoral pair — high on body, short and rounded. Pelvic pair — forward almost under pectorals. Anal — large with 3 spines at front. Tail — broad, forked, tips rounded.

Spawning: July-August. Water temperature ± 24°C. Male builds nests and guards eggs.

Life Span: Maturity at 2-4 years. Maximum age 8 years.

Food: Insects, other invertebrates, small fish.

Habitat: Weedy streams, ponds and bays. Shallow, clear, almost still water.

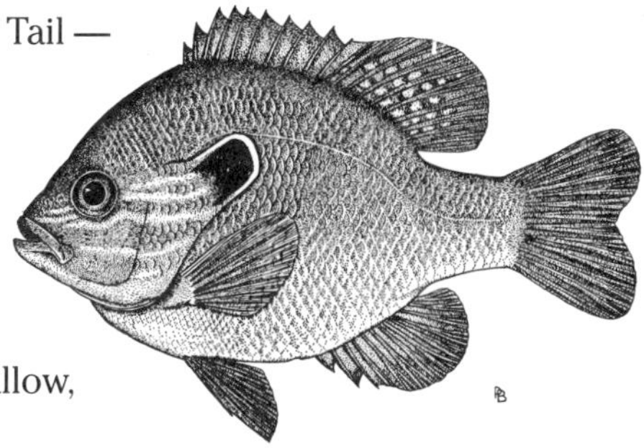

7. Golden Redhorse G5 NAR S3

Moxostoma erythrurum **Rafinesque**

8. Greater Redhorse G3 S3

Moxostoma valenciennesi **Jordan**

These two Redhorses or Mullets are closely related large-sized members of the Sucker family, with thick robust bodies and large scales. As the family name implies, they feed by sucking up invertebrates from the bottom sediments in the shallow waters they inhabit. The origin of the English name is something of a mystery, unless one can be persuaded that the snout is somewhat similar to that of a horse! There is usually some part that is reddish tinged. They are certainly very blunt nosed and with a rather smooth oval outline. These are both fish of stream habitat, although they may be found in Lake Huron near the mouth of the streams. The Golden Redhorse is, on average (32 cm), slightly smaller than the Greater Redhorse (average 42 cm). The Golden Redhorse, which ranges across the eastern and central United States, is at the northern edge of its range in Lake Huron and is only found in the waters of Bruce County. The Greater Redhorse has a similar distribution but its range extends somewhat further north to the Spanish River near Sudbury, so it is also recorded for Grey County, i.e. from the vicinity of Georgian Bay. The Golden Redhorse is more silt tolerant than the Greater. This would cause the latter to be more vulnerable if there is degradation of the water quality.

Golden Redhorse Description:

Length and Shape: 28-46 cm. Laterally compressed. Little arch to back. Tapering to rear. Head squarish, slightly less than 20% of body length. Eye small. Mouth large, below, protrusible (sucker).

Colour: Grey to bronze or olive, golden sheen in adults, sides paler. Silver to milky-white below.

Fins: Dorsal — large, central, slate grey. Pectoral pair — low on body, long, bluntly pointed. White to pale orange (also pelvic and anal fins). Pelvic pair — central, low, square-ended. Anal — long, rounded in female, more pointed in the male.
Tail — large, forked with sharp points, slate-grey. All fins may be orange to red orange.

Spawning: Spring. In riffles in stream. No parental care.

Life Span: Mature ± 4 years. Maximum 7-8 years.

Food: Insect larvae, worms, molluscs.

Habitat: Slow moving streams with soft bottoms, relatively clear water. Generally sedentary, remaining in one pool while food is adequate.

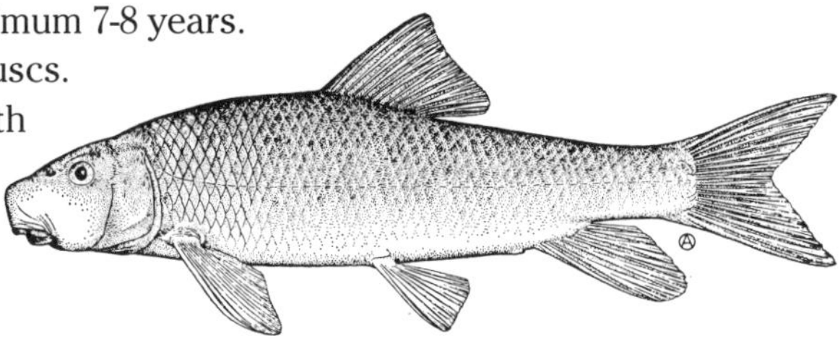

Greater Redhorse Description:

Length and Shape: 38-50 cm.
The largest of the Redhorse Suckers. Somewhat laterally compressed. Back with a shallow arch. Head massive, long (20% of body length). Eye small. Mouth below, protrusible, lips thick.
Colour: Olive-green with bronze sheen. Sides paler, more golden. Below golden or milk-white.
Fins: Dorsal — 1 large, central, deep red. Pectoral pair — low on body, long fairly wide, whitish to olive behind. Pelvic pair — low, central, fairly wide, whitish to pale red behind. Anal — long, deep red. Tail — long, wide, blunt pointed, deep red.
Spawning: May-June. In fairly rapid streams.
Life Span: Unknown.
Food: Immature insects, worms, molluscs.
Habitat: Large streams with clear water. Bottom with clean sand, gravel or boulders. Probably also in lakes near stream mouths.

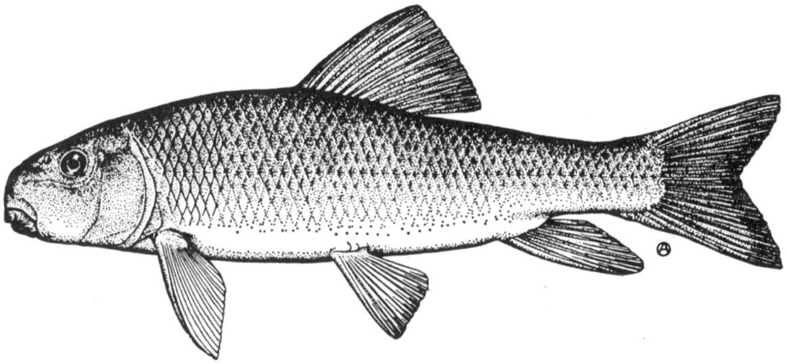

REPTILES

Reptiles are cold blooded vertebrates which may live on land or in aquatic environments. In this area winter temperatures fall below the level at which these creatures are able to function. It is, therefore, necessary for them to find some suitable form of shelter in the autumn, so that they can hibernate through the cold winter months. There is a large winter die-off of newly hatched individuals who do not survive the cold, and in an especially severe winter many of the older ones also will die. Most reptiles can be described as predators, although their prey may range from insects to quite large mammals. In this area we have twelve snakes, one lizard and six turtles. Of this number, three snakes and two turtles are considered to be threatened or vulnerable in Ontario. The Eastern Hognose Snake, is also ranked S3 and vulnerable in Ontario but has not been reported in this area since 1972. The Five-lined Skink (S3) has been reported once. Habitat destruction is the major threat to these species. Many reptiles either live in water or inhabit wetlands so they are vulnerable to all the practices which damage or destroy these habitats. Farming and logging also destroyed natural land areas with which these species were associated, and roads fragmented their territories. With respect to snakes, deeply seated ancestral memories of snakes as evil and dangerous, and the assumption that all of them are poisonous, make them vulnerable to destruction by many people. In actual fact, snakes are often very beneficial as they keep unwanted species such as insects and mice under control. Certainly, snakes have evolved a remarkable sensitivity to vibration which assists them in locating their prey, but also serves as an early warning signal of danger. Consequently, they most often sense you before you see them, which makes them remarkably elusive even in places where they are plentiful. Turtles, which are of much more ancient origin, have evolved their own form of armour plate into which they can retreat. They are most at risk in the breeding season when they tend to cross roads to get to their nesting sites. Most reptiles lay eggs which have a pliable parchment-like shell. At this stage they are very vulnerable to predation. Fertilization is internal, and some snakes have evolved one step further so that the embryo is retained in the female's body and is born at a much more developed stage. The Eastern Massasauga Rattlesnake is an example of such a viviparous species, and it is also the only poisonous one occurring here. Some reptile females also have the ability to store sperm for up to a few years, so they are able to breed under suitable conditions even if a male is not available, which may be the case after a very cold winter. Another interesting feature of reptiles is that most of them continue to grow throughout their life span so that, within the limits of a given species, the largest specimens are the oldest. For this reason, snakes periodically shed their skin when it becomes too tight. In turtles, of course, the shell grows in order to give more space.

Important: None of the species listed in this section may be harvested. They are protected as non-game reptiles (amphibians have similar protection) under the Fish and Wildlife Conservation Act (1999). For educational puposes only it is permitted to have one specimen of any given species in captivity. It is against the law to possess more than one live specimen.

1. Spotted Turtle **G5 SC S3 VUL**

(*Clemmys guttata*)

This small turtle is easily recognized by its black colour and scattered bright yellow spots. The only species with which it can be confused is the Blanding's Turtle, but that is much larger (15-25 cm), brownish rather than black, and with yellowish flecks rather than spots. The Spotted Turtle is found in eastern North America with a range extending from Michigan through southern Ontario to New England and south through Ohio, Pennsylvania and the Atlantic coastal states to Georgia with a couple of isolated stations in Florida. The population in Bruce, Grey and Simcoe Counties is somewhat isolated from the main range. This is a species of cool, shallow waters. It is the first turtle species to emerge in the spring and they may be seen basking in a sheltered spot as early as March, even though the wind may be freezing. They do not take kindly to heat, and actually go into a form of summer hibernation, known as aestivation. If the temperature in their habitat becomes too high, they will bury themselves in the vegetation at the bottom and not emerge again until temperatures drop in the fall.

Local Distribution: Wetland areas of MacGregor Point Provincial Park. Similar habitats along the west shores of the Bruce Peninsula such as Petrel Point. Bruce Peninsula National Park 1990.

Habitat: Cool woodland ponds, fens, marshes, sloughs along the shorelines of large lakes and rivers. (Peat bogs in eastern Ontario).

DESCRIPTION

Length: 8-12 cm (maximum 12.7 cm)

Colour: Shell — black with yellow spots. Young turtles have fewer spots. Old ones have faded spots. Head and legs — black and spotted. Males with black lips. Females with pinkish lips.

Food: Mostly aquatic insects. Also snails, fish, tadpoles and aquatic plants.

Seasonal Behaviour:

Emergence from hibernation: (March) April.

Mating: May.

Egg Laying: June, 1-8 eggs, on sandy beaches or well drained soil.

Hatching: September.

Hibernation: October.

2. Wood Turtle **G4 SC S2 VUL**

(Clemmys insculpta)

[Protected by the convention on International Trade in Endangered Species, as well as the Ontario Fish and Game Act.]

The Wood Turtle is of medium size and associated with rivers and streams. In the summer it spends much of its time foraging for food in the open woods that border these waterways. This turtle has a similar range to the Spotted Turtle in Eastern North America, but extends further west into Wisconsin and Minnesota and only as far south as Maryland. The Wood Turtle spends part of its life on land. It has a ridged, sculptured shell and it walks with its shell held above the ground. The Snapping Turtle also has a ridged shell, but it tends to be much larger and it is mainly aquatic. When it does emerge from the water it drags its shell on the ground. To their detriment, wood turtles are frequently collected as pets. They are believed to be very intelligent because they will not walk off the edge of a table, whereas a Painted Turtle will go straight over the edge and plunge to the floor. This is actually an innate instinct which evolved because Wood Turtles are partially terrestrial and frequently walk on logs and rocks. Diving off these could be fatal. The Painted Turtle is aquatic and when it emerges and basks on a log it is accustomed to diving into the river when disturbed. One of the more engaging habits developed by the Wood Turtle is stomping on the ground, the vibrations causing worms to emerge, and the turtle eats them. A recent study (University of Guelph) indicated 400 individuals in the Lake Huron population. It is now estimated that there has already been a 50% reduction due to illegal collecting for the pet trade. In order to facilitate an increase in population number, the Huron Fringe Field Naturalists with the assistance of the MNR have been protecting Wood Turtle nests along a local river. In June they mount a watch for nesting turtles. They then cover the nests that they discover with wire mesh to protect them from predators. In August, just before the eggs hatch, they remove the mesh. They also create nesting sites by clearing brush and spreading sand. Suitable undisturbed, wooded river bank habitat is very important for the survival of this species.

Local Distribution: Only reported from southern Bruce County.

Habitat: Rivers and streams with sandy or rocky bottoms. Open woodlands or willow thickets along the banks.

DESCRIPTION

Length: 13-23 cm.

Colour: Brown with sculptured plates composed of incremental annual growth rings.

Food: Snails, slugs, insects, molluscs, frogs, tadpoles, flowers, fruits such as berries, willow leaves.

Seasonal Behaviour:
Emergence from hibernation: April.
Mating: May. In the water.
Egg Laying: June. 4-18 eggs. Along river banks.
Hatching: Late August. Young Wood Turtles spend most of their time in the water.
Hibernation: October. In gravel bottomed rivers. Wedged under tree roots or in muskrat burrows.

3. Eastern Fox Snake G3 THR S3 THR

(*Elaphe gloydii*)
[*Elaphe vulpina* var. *gloydii*]

This is the largest snake recorded in this area and has been known to reach a length of 179 cm, although it is usually rather smaller but seldom less than 1 metre at maturity. The head ranges from light brown to reddish which often causes the inexperienced on first sight to believe that they are seeing a Copperhead. The overall body colour is yellow with dark brown or black blotches. Unfortunately, the Fox Snake has a habit of vibrating its tail laterally in dry leaves, which, with the slight similarity of patterning, may cause it to be mistaken for a rattlesnake by the uninitiated. This eastern variety of the Fox Snake was probably never common in the two counties because of the lack of suitable habitat. It is a creature of open areas close to the shoreline i.e. beaches, sandspits, fields and marshy places. Its range is now limited to eastern Michigan and north-central Ohio, where there are remnant populations; the remaining 65 percent of the world distribution is found in southern Ontario, along the east shore of Georgian Bay to just north of Parry Sound and in the Long Point area and the extreme southwest corner of the province. It is recognised by farmers as a good "mouser." This should have endeared it to the early settlers who were overrun with deer mice that stole food off the table!

Local Distribution: There is a record from the Bruce Peninsula National Park area in 1974. There have been no other confirmed reports since 1950. It possibly could be confused with the Northern Milk Snake which is similarly patterned but has a pale "Y" shaped marking on its head. It also vibrates its tail when threatened.

Habitat: Unforested areas adjacent to the marshes and shores of Lake Huron and Georgian Bay — beaches, sandspits, exposed rocks, fields.

DESCRIPTION

Length: ± 91 to 137 cm. Head oval. Tail pointed.

Colour: Yellowish. Dark brown to black blotches on back. Smaller blotches on sides alternating with back blotches. Yellow belly chequered with black. Young are lighter in colour.

Food: Small mammals — especially deer mice and meadow voles. Frogs. Also bird's eggs and nestlings — reached by climbing. Larger prey are killed by wrapping coils around the body and constricting them.

Seasonal Behaviour:

Emergence from Hibernation: April.

Mating: May and early June.

Egg Laying: June and July 15-20 white eggs, 4 cm long, rubbery.

Under dead wood on beaches, and in rotting logs or stumps. Also in manure and sawdust piles.

Hatching:

August September. Length ± 30 cm.

Hibernation: October to early November.

4. Queen Snake **G5 THR S2 THR**

(Regina septemvittata)

[*Regina* — Queen; *septem* — seven; *vittata* — bands]

The Queen Snake is a medium sized water snake. It can be mistaken for a Garter Snake because of the broad yellow stripe on each side, standing out against the olive-green to brown of the background. In Garter Snakes the background is black. In juveniles and young adults three additional dark stripes can be seen along the back and two on each side — "queen with seven bands." These snakes are scarce and elusive, and almost always found in or near water. They are agile swimmers and divers, and often hide during the day under rocks in the stream bed or in the debris at the edge. In the spring and fall on sunny days they will bask among the low bushes and tree trunks beside the stream. As these bushes are often willows, they are sometimes called Willow Snakes.

Queen Snakes are widely distributed in the United States east of the Mississippi River. They are on the edge of their range in Canada. They are only found in south-western Ontario in streams running into Lake Erie and Lake Huron, the Bruce Peninsula is their northern limit. The main reason for their scarcity in this area is climatic; they are better adapted to warmer areas. A large part of their diet consists of young or newly moulted crayfish which occur in the clear, rocky streams inhabited by this snake. They are, therefore, susceptible to pollution of these waters, and to increased run-off and siltation which eliminate the crayfish population.

Local Distribution: In 1989 Fred Schuler of the Canadian Museum of Nature reported this snake in the Bruce Peninsula National Park (St. Edmunds Twp.). There is a report slightly earlier from Sauble Falls, and historic records from the Johnson Harbour area.

Habitat: Clear spring-fed streams with moderate to fast currents and rocky bottoms.

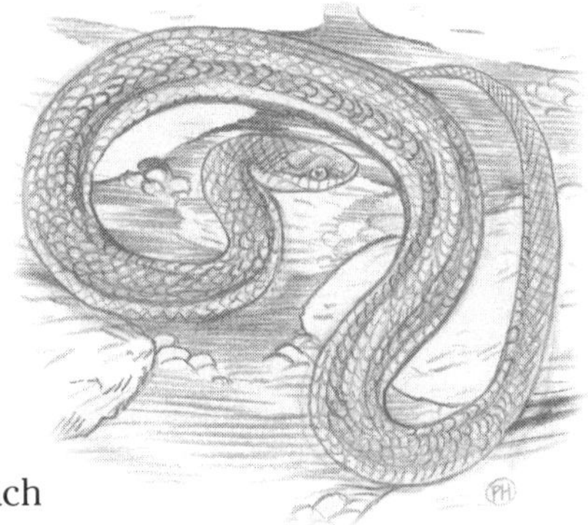

DESCRIPTION

Length: 30-60 cm

Colour: Olive-green to brown. Off-white belly extending through upper lips. Broad yellow stripe on each side.

When young — 3 dark stripes on back and 2 on each side.

Food: Freshly moulted crayfish. Also frogs, newts, tadpoles, minnows, snails and fairy shrimp.

Seasonal Behaviour:

Emergence from hibernation: May

Mating: Spring or fall. (If in the fall, embryo development is suspended until spring).

Birth: Live young (160 mm long) born in August or September. Between 5 and 23 — average 13.

Hibernation: September to May.

5. Eastern Massasauga Rattlesnake

(*Sistrurus catenatus catenatus*) **G3G4 THR S3 THR**

This snake is a member of the pit viper family which have a heat-sensing pit between each eye and nostril. These enable them to sense the presence of small warm-blooded mammals on which they prey. Their venom is a specialized digestive enzyme, which disrupts blood flow and prevents blood clotting. It is injected through the snake's retractable hollow fangs. Exposure to rattlesnake venom is immediately lethal, to a frog or mouse. The purpose of the venom is to kill small prey, not humans. However, the snake should not be handled without proper precautions. In the very unlikely event of being bitten it is important to get to the nearest hospital immediately. Local hospitals have antivenin (antivenom) available. Often no poison will have been injected as the snake was biting defensively. Only two people are known to have died from snake bite in Ontario, and neither received appropriate or timely medical treatment. As a comparison, on average, six people are killed by lightning every year. This is the only Ontario snake that has cat-like eyes with vertical slits. They are most likely to be visible near their hibernacula in cold weather in spring or fall when they will bask in the sun but, because of the temperature, they do not react quickly. In the summer, they are sometimes found in drier, rocky upland areas. This is a rather heavy-bodied snake with a triangular head and a blunt tail which ends in a rattle. When they sense your presence, they will raise their tail and rattle a warning which sounds like an insect-like buzz, and then slide away. Non-venomous snakes that produce a rattle-like sound do so by vibrating their tail against something else. Rattlesnakes were once widespread in central and southern Ontario. Plowing of prairie areas and draining of wetlands has resulted in habitat loss. In this area, extensive cottage development, farming, and persecution of this species because of the unwarranted fear it generates, has resulted in a reduced population. The staff of Bruce Peninsula National Park are monitoring rattlesnakes on the upper Bruce Peninsula with capture-release programs and radio telemetry. In other areas controlled burns are being used to regenerate the prairie habitat favoured by this snake. Education programs are encouraging people to change their attitudes. A National Recovery Team has been established to address the conservation needs of this rare and declining animal.

Local Distribution and Habitat: Fens, beaver meadows, shorelines in spring and fall. In summer they may move through upland mixed and coniferous forests. There have been no records in Grey-Bruce of this species off the Bruce peninsula for many years.

DESCRIPTION

Length: 51-76 cm. Head triangular. Tail blunt. Heavy body.

Colour: Grey or brown with saddle-shaped black blotches along the back. Smaller blotches on each side. Belly black. Dark brown rattle at end of tail. Young with patterns but paler.

Food: Mice, chipmunks, and other small rodents, small birds, frogs, snakes.

Seasonal Behaviour:
Emergence from Hibernation: Late April – early May.
Mating: Late summer to early September.
Birth: July or August. 7-8 live young ± 20 cm long. Fully equipped with fangs and venom.
Hibernation: October. Associated with groundwater (not just in wetlands).

Important Note: Residents who find a rattlesnake on their property and are concerned can contact their local Ministry of Natural Resources office or the National Park staff for advise, as appropriate to where they are situated. Do not kill the snake!

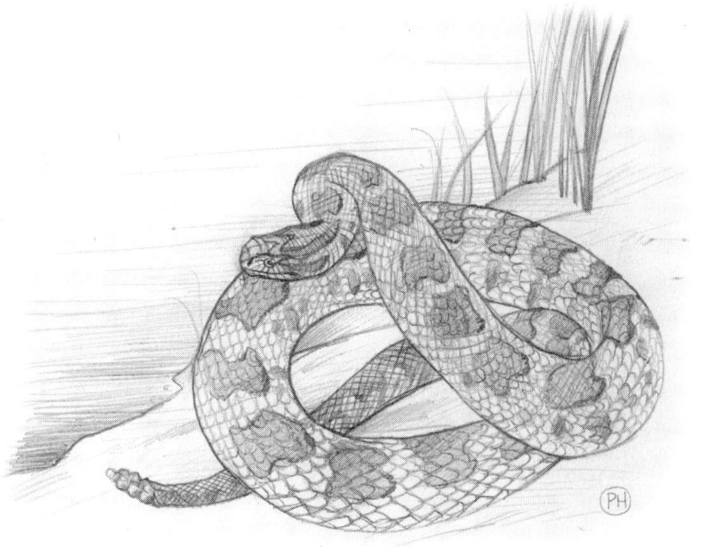

References:

Behler, J.L. and King, F.W. 1979. **The Audubon Society Field Guide to North American Reptiles and Amphibians.** A. Knopf, New York.

Conant, R. and T.J.Collins. 1991. **Field Guide to Reptiles and Amphibians of Eastern and Central North America**. Third edition. Houghton Mifflin Co., Boston

Cook, F.R. 1984. **Introduction to Canadian Reptiles and Amphibians.** Nat. Mus. of Canada.

Harding, J.H. 1997. **Amphibians and Reptiles of the Great Lakes Region.** University of Michigan Press, Ann Arbor.

Johnson, Bob. 1989. **Familiar Amphibians and Reptiles of Ontario.** Natural Heritage/Natural History Inc., Toronto.

Logier, E.B.S. 1958. **The Snakes of Ontario.** University of Toronto Press.

BIRDS

Editorial privilege has been used to select only information deemed relevant to the purposes of the present publication. That is, to describe the **breeding** (B) status of rare, endangered and vulnerable species in this area and encourage an interest in their preservation among the widest number of people. The species included have involved arbitrary decisions made necessary by a number of factors. These include extralimital occurrences of some species that may not be uncommon throughout their range (e.g. Great Egret); the lack of a large amount of suitable habitat (e.g. marshes for rails and allies); some rare breeding species in our area, whose populations are poorly known (e.g. Northern Parula, Cape May Warbler); those species presently showing an increase in Ontario but uncommon in our area (e.g. Bald Eagle, Peregrine Falcon); and endangered species (e.g. Piping Plover) which are almost extirpated from Ontario, and no longer breed in our area.

The Passenger Pigeon, *Ectopistes migratorius* is the only Grey-Bruce breeding bird species that can claim in historical times to have obtained the COSEWIC status: Extinct. There are two published records of dated nestings (Mitchell 1935). The first was in Amabel Twp. on 28 April 1876; and the second in Lot 13, Conc. D, Amabel Twp. on 24 May 1878. This latter nesting involved a colony of 'hundreds of thousands' of birds. It has been estimated that at the time of the discovery of North America the population of Passenger Pigeons was 3-5 billion birds, and that it represented 25 to 40 percent of the total bird population of the present United States. In a Michigan nesting colony 25,000 pigeons a day were killed for the market in a month by netters. The last Passenger Pigeon died in a Cincinatti zoo in 1914.

There are two lists. The main list has 14 species whose inclusion is felt to be of primary importance, and a supplementary list of 10 species whose status in Grey-Bruce also warrants consideration. The lists are in the taxonomic order proposed in the seventh edition (AOU 1998) of the American Ornithologists' Union Checklist. References to North America refer to the area designated by the American Birding Association (ABA), and is that area north of Mexico which includes the United States and Canada but excludes Greenland, Hawaii and islands in the Atlantic and Caribbean that are more than 200 miles from the mainland.

All of these birds breed or have bred in Grey and Bruce, and this fact plus their status were the main reasons for their inclusion. There are obviously a number of non-breeding migrants and accidental and extralimital species which have occurred in the area and have not been included here. Since active nests are the best proof of breeding, the map symbols in the main list are shown in the township units where nests have been found, and in a few cases where breeding is known to have occurred. The nesting data was extracted from the huge database of the Ontario Nest Records Scheme (ONRS) at the Royal Ontario Museum in Toronto and the various literature references resulting from the summarization of those nest records (Peck and James 1983, 1987, 1993a, 1993b, 1994, 1997, 1998a, 1998b, 1999). In some areas there is evidence that breeding is still taking place, but this has not been confirmed by nest sightings. In addition, some of these birds may be seen moving through the area, even though they are not actually breeding here. The current Canadian status of these species in the main list is that provided by the Committee on the Status of Endangered Wildlife in Canada (COSEWIC) up to 2001 and is placed after the G rank.

1. Least Bittern G5 SC S3B VUL

Ixobrychus exilis

The Least Bittern is a member of the family Ardeidae which includes bitterns, herons and egrets. There are 17 species occurring in North America. These wading birds have elongated necks that form a kinked S-shape, and they reside mainly in wetlands where they feed on amphibians, fish, insects and other invertebrates. The Least Bittern is the smallest North American heron and one of the most secretive. It dwells in areas of dense marsh vegetation, and not being a strong flier, it seldom flushes and thus is not often seen. If disturbed it may 'freeze' in a typical hiding pose with its neck extended and bill pointed upward. At this point it closely resembles the dead reeds nearby, and should there be a breeze it may even sway slightly in unison with the movement of the vegetation. This characteristic hiding pose is even adopted by the nestlings when disturbed.

Local Status: The Least Bittern is a declining species whose population in Grey and Bruce is influenced by a limited amount of suitable marsh breeding habitat.

Nest records: *Bruce*: 2 nests (1929, 1932) from Silver Lake, near Kinloss, Kinloss Twp.; 2 nests (1938, 1973) from Sky and Isaac Lakes, Albemarle Twp.; and 1 nest (1976) on the Rankin River, Amabel Twp.

Grey: current breeding evidence has been reported from Keppel Twp.

IDENTIFICATION: Length 28-36cm; wingspan 43cm. Male: crown and back black; underparts and prominent inner wing patches buff; bill pale, legs dull greenish. Female: crown and back chestnut. Immature: resembles female, but has more marked streaking on back and breast.

Voice: a low, descending *poopoopoo;* call includes a series of *kek* notes.

Food: mainly small fish; also aquatic insects, other invertebrates, frogs, tadpoles, molluscs.

Habitat: during the breeding season is usually found in dense freshwater marshes of reeds, cattails and other vegetation.

Breeding: Nests are supported by living and dead standing rushes, grasses and occasionally shrubs, and are positioned over shallow water at heights of 5 to 120cm.

Movements: winters from southern USA to northern South America. Average spring arrival in mid-May.

2. Great Egret

G5 Not listed S2B

Ardea alba

[Casmerodius albus]

The Great Egret is a member of the heron family and has a world-wide distribution. It breeds from coast to coast across the southern United States and north to southern Canada in the east. It inhabits both coastal and inland wetlands of all kinds. It is the second-largest heron in North America.

Local Status: This colonial species is at the northern limit of its breeding range in Ontario, although it is locally common throughout much of the United States. In Grey and Bruce it has only been recorded nesting on Chantry Island, near Southampton in Saugeen Twp., Bruce County. Nest records have been submitted in 1990 and 1999 from a mixed colony with Great Blue Herons and Double-crested Cormorants on Chantry Island. It has probably bred on Chantry Island each year since 1990. It is worth noting that Ontario's most northerly nesting location is nearby, on Nottawasaga Island, near Collingwood, Simcoe County where the species has nested annually since 1985. These birds are frequently seen feeding in Grey County wetlands such as Little Germany.

IDENTIFICATION: length 80-104cm; wingspan 140-170cm .
Both sexes white with yellow bill and black legs. Breeding adults have long plumes on the back.
Immature: as adult but yellow bill has black tip.
Voice: very low, deep *kroow,* and other low croaks.
Food: fish, amphibians, snakes, aquatic insects and crustaceans; also small birds and mammals.
Habitat: wetlands, including marshes, flood plains, shores, swamps, and mudflats.
Breeding: usually in colonies, sometimes mixed, and in our area on islands in large lakes and rivers. Nests are bulky platforms of sticks and twigs situated over or near water in trees, shrubs and vines, and at heights from 0.6 to 11m.
Movements: winters from southern California across southern United States, and south along coast from New Jersey through South America. Arrives in our area in April.

3. Black-crowned Night-Heron G5 S3B

Nycticorax nycticorax

A cosmopolitan species, the Black-crowned Night-Heron is a member of the family Ardeidae, and breeds on every continent except Australia and Antarctica. In the Western Hemisphere it occurs from southern Canada south to southern South America. This chunky heron becomes more active at dusk and does most of its foraging from then to dawn.

Local Status: In Ontario this colonial species is toward the northern limit of its breeding range, although it is not uncommon south of Canada in the United States.

Nest Records: *Bruce*: nesting colonies (1967, 1968, 1989) reported from the Fishing Islands, Albemarle Twp. (Kirk and Peck 1984); and from Chantry Island, Saugeen Twp. in 6 years between 1975 and 1989. It is reported that it has bred each year on Chantry Island to 1999, although no nest records are yet available. Two nearby nesting locations are James Island, Manitoulin District (1994), and annually on Nottawasaga Island, Georgian Bay in Simcoe County.

IDENTIFICATION: length 56-65cm; wingspan 105-112cm.
Adults have a black cap and back, pale gray wings, whitish underparts, a red eye, black bill and yellow legs.
Immature: dull gray-brown with numerous light spots on upper parts, and streaks on underparts.
Voice: a flat, barking *quok* or *quark*.
Food: varied diet consisting of fish, frogs, tadpoles, lizards, insects, crustaceans, molluscs, eggs and chicks of other species.
Habitat: found in a wide variety of aquatic habitats in both fresh and salt water.
Breeding: usually in colonies both homogeneous and mixed. Nests are platforms of twigs and sticks in shrubs and trees, at usual heights of from 1.8 to 3.7m. Occasionally unelevated nests are placed on mats of vegetation in marshes.
Movements: winters south throughout the breeding range. Arrives in our area from mid-April to early May.

4. Bald Eagle **G4 NAR S4B END**

Haliaeetus leucocephalus

The problem populations in southwestern Ontario are currently recovering, and there has been an encouraging growth in successful fledging. In 2000, there were 22 active nests reported in southern Ontario.

The Bald Eagle is one of four eagles in North America, all belonging to the family Accipitridae which also includes the osprey, kites, harriers, and hawks. It breeds in most of Canada except the far north and the plains region, and south on both coasts to the southern states and only locally in the interior. It is absent from most of the prairie and plains regions.

Local Status: In Grey-Bruce no nests were reported after 1966 for 26 years, a time period which corresponded with the decline of this fish-eating species along the Great Lakes in the 1960s due to the effects of DDT. Although a number winter in our area, mostly at Lake Huron near the Bruce Nuclear Power Station at Douglas Point, no eagles have nested successfully in Grey and Bruce in the last several years.

Nest Records: *Bruce*: a nest (1930, 1938) at Howdenvale, Albemarle Twp.; this nest was also reported in use for 10 to 12 years prior to 1930. At Sydney Bay Bluffs, Albemarle Twp., three nests were intermittently used in the years between 1948 and 1952. At Stokes Bay, Eastnor Twp., nest records were submitted for 1937, 1938, and 1960. In Amabel Twp. near Wiarton, a nest with 2 eggs was reported in 1938.

Grey: at Meaford Tank Range, St. Vincent Twp. nesting was reported from 1962 to 1966, and in the 1990s (1992, 1993, 1994, 1995, 1996) also in 2000 and 2001. This nest is now the only site in Grey and Bruce to have been recently used by this species.

IDENTIFICATION: length 71-96cm; wingspan 168-244. As is true of most of the birds of prey, the female is often considerably larger than the male. Adults are chocolate-brown with a white head and tail; bill and feet are yellow. Immature: variable mottled shadings of light tan to brownish-black, becoming lighter with successive molts until adult plumage is attained at 3 to 4 years; bill black.
Voice: a rather weak, chirping whistle.
Food: variable and includes fish, carrion, birds, reptiles, mammals, and invertebrates.
Habitat: generally prefers areas bordering water (coasts, lakes, rivers).

Breeding: in Ontario breeds usually in woodlands; the majority of nests were on wooded islands in lakes and large rivers. Nests are huge, cup-shaped platforms with flat tops and are built of dead branches and sticks, and lined with grasses and other finer material. Nests are in trees at heights averaging 15 to 21m.
Movements: winters south throughout the breeding range.

5. Red-shouldered Hawk G5 SC S4B VUL

Buteo lineatus

The Red-shouldered Hawk is a member of the genus *Buteo*, family Accipitridae, with 10 species in North America. The buteos are characterized by long, broad wings, robust bodies, short to medium tails, and they soar regularly. A denizen of wet woodlands where it breeds and hunts, the Red-shouldered Hawk is seen less often than, and its behaviour is necessarily somewhat different from that of, a more common relative, the Red-tailed Hawk. The latter frequents more open terrain. The Red-shouldered Hawk breeds from southeastern Canada south through the central and eastern United States, and in California.

Local Status: probably still breeds in small numbers each year in our area, although no nests have been reported since 1996.

Nest Records: *Bruce*: a nest (1933) at Lucknow, Kinloss Twp.; a nest with 2 young (1938) at Chalk Lake (name & location unknown); a nest with 3 young (1985) near Oliphant, Amabel Twp.; and an active nest (1994) at Cameron Lake, St. Edmunds Twp.

Grey: 11 nests (1969-1996) in Holland Twp.; and an active nest (1995) in the Glen Conservation Area, Keppel Twp.

IDENTIFICATION: length 43-61cm; wingspan 90-127cm. A medium-sized hawk; the female is larger than the male. Adults are tan to rufous, with black and white patterned wings and tail; reddish shoulders and pale crescents at base of primaries (in flight) are the best field marks. Immature: brown to tannish with streaked underparts.
Voice: repeated cries, *keeyar, keeyar, keeyar;* call often imitated by Blue Jay.
Food: varied; small mammals, reptiles, amphibians, birds, and insects.
Habitat: mature, lowland woodlands with nearby water.
Breeding: invariably nests in trees, usually living; and nests are frail to bulky platforms of sticks and twigs, commonly lined with green coniferous sprigs. Nest heights averaged 11.5 to 15.5m.
Movements: winters through the breeding range but generally south of Canada. Average spring arrival is early to mid-April.

6. Peregrine Falcon G4 SC S2S3B END

Falco peregrinus

This status designation applies to the race or subspecies breeding in our area (*Falco peregrinus anatum*). The current provincial increase of this species is due largely to the re-introduction of captive-reared birds at 'hacking' sites in the province.

In North America the genus *Falco* in the family Falconidae is represented by eight species, two of which are accidental. The true falcons have long, pointed wings and medium to long tails. Like other birds of prey they have hooked beaks and curved, sharp talons. The peregrine is one of the swiftest fliers in the bird world and has been clocked at 280 km per hour in a dive or stoop, while in pursuit of prey. It breeds locally throughout North America with the largest concentrations found in arctic regions, the Rockies and on the coasts. The race (*anatum*) in our area was all but exterminated by the use of pesticides, but is presently being reintroduced.

Local Status: a former breeder on the Niagara Escarpment in Grey-Bruce, no nests have been noted here since 1940, until a nest was recently found in Bruce County in 2000.

Nest Records: *Bruce:* an active nest (1915) with adults present at Cape Croker (Williams, 1942); an active nest reported (2000) from Eastnor Twp. is the first new nesting in 60 years in Grey-Bruce.

Grey: six nest records (1929-1938) from a site at 'Metcalfe's Rock' near Redwing, Collingwood Twp.; and two nest records (1937, 1940) from Eugenia Falls, Artemesia Twp.

IDENTIFICATION: length 35-50cm; wingspan 80-120cm.

Adults: large, dark falcon with wide, dark mustache mark, dark head, upper parts and underwing; white breast contrasts with darker marked belly; cere, legs and eye-ring are yellow. Immatures: upper parts have tones of black to pale brown, and streaked underparts.

Voice: a harsh *kyah, kyah, kyah* (near nest).

Food: chiefly birds; occasionally mammals, reptiles, and insects.

Habitat: variable throughout year; open country, cliffs, and sometimes cities during breeding season.

Breeding: nests on cliffs overlooking lakes, on island cliffs in lakes and rivers, and occasionally on cliffs not adjacent to water. In recent decades some nests have been on ledges of tall buildings in urban areas. Nests are mere depressions in substrate.

Movements: winters from central and southern states south to central and South America. Average spring arrival is early to mid-April.

7. Piping Plover **G3 END S1B END**

Charadrius melodus

Plovers are shorebirds belonging to the family Charadriidae, with 17 species in North America, eight of them accidentals. One of the commonest and best-known members of the family is the Killdeer. The Piping Plover breeds in the Great Plains, the Great Lakes and on the Atlantic coast from Newfoundland to North Carolina. It almost became extinct around 1900 due to excessive hunting. After a population recovery it is again in decline throughout its range, now because of increased recreational use of nesting beaches and development in its winter habitat. It is all but extirpated in Ontario, with one or two pairs still nesting at Lake of the Woods, in Rainy River District. In 1996 there were only 16 pairs left in the Great Lakes area.

Local Status: The Piping Plover last nested in our area in 1972.

Nest Records: *Bruce:* 5 nest records (1966-1972) from the beaches near Oliphant, Amabel Twp.

IDENTIFICATION: length 17-18cm; wingspan 48cm. A small, chubby plover with pale coloration tending to camouflage it on its sandy breeding habitat. Adult (breeding): pale, beige upperparts and white underparts; dark, narrow band on breast and head; orange bill with black tip; legs orange. Immature: like nonbreeding adult, lacking black bands; bill black.

Voice: a whistled *peep* or *peep-loh*.

Food: aquatic insects, crustaceans and molluscs.

Habitat: sand and gravel beaches and mud-flats.

Breeding: in Ontario breeds on sand and gravel beaches on shores of large lakes and rivers. Nests were shallow scapes or depressions in sand and gravel, either unlined or lined with small flat pebbles and occasionally grasses. The eggs are cryptically-coloured.

Movements: winters mainly on the Atlantic-Gulf coast from North Carolina south to Florida and eastern Texas; less commonly south to the Caribbean and Mexico. Arrives in spring in mid-May.

8. Caspian Tern **G5 NAR S3B**

Sterna caspia

Since its vulnerable designation in 1978, its numbers have increased, resulting in its current delisting.

Terns along with gulls, skuas and skimmers are members of the family Laridae. There are 19 species of terns in North America. Terns are slender, graceful fliers with long, pointed wings, and are generally smaller than gulls. The Caspian Tern is a size exception, and is the largest of the world's terns. In size it is midway between Ring-billed and Herring Gulls. A cosmopolitan species, the Caspian Tern is common along both coasts of the United States and Canada, and is found locally inland, mainly on islands in large water bodies.

Local Status: There are presently 11 colonies in Ontario, one of which is in our area.

Nest Records: *Bruce*: a nest record (1963) from an island in the Fishing Islands, Amabel Twp.; and records of the colony (1980, 1981, 1991) on Halfmoon Island, Georgian Bay, St. Edmunds Twp. This latter colony ranged from 150 to 250 pairs in those years. There have been no more recent records submitted for this colony.

IDENTIFICATION: length 48-56cm; wingspan 127-140cm. A large, gull-like tern with a thick, blood-red bill with variable black tip. Adult (breeding): black cap, gray upper parts, white underparts; black under-surface of primaries conspicuous in flight; legs black. Immature: gray above with brown bars; crown mostly white.
Voice: a deep, harsh *kahhrrrr*, and repeated *kaks*.
Food: mainly small to medium sized fish; occasionally eggs and young of other birds; it plunge-dives for fish.
Habitat: locally along coasts, and as well on large, inland lakes.
Breeding: primarily a colonial species, nesting sites in Ontario were often shared with Ring-billed and Herring Gulls and Common Terns. With one exception (a manmade landspit), all nesting sites in Ontario were on shoals and islands in large lakes. Nests are depressions in the substrate, some unlined, but most lined with grasses, pebbles, fish bones, twigs, and feathers. Average-sized clutch is 2 eggs.
Movements: winters in the Americas mainly in coastal areas of the southern United States, south to Mexico, Central America and northern South America. Arrives in our area near the end of April.

9. Black Tern **G4 NAR S3B VUL**

Chlidonias niger

The declining numbers of this species in recent decades warrant its being kept under careful scrutiny. The main reason for its decline in certain areas apparently relates to the disappearance of wetland habitat.

The Black Tern is a small, graceful bird with a buoyant, swallow-like flight and in its adult breeding plumage is unmistakable. In North America, in summer, it is a typical bird of freshwater marshes, but in winter it becomes a seabird, found along tropical coasts.

Local Status: A relative lack of suitable marsh breeding habitat has restricted the numbers of this tern in Grey-Bruce.

Nest Records: *Bruce*: a nest record (1982) from Rankin River, Amabel Twp.; and 11 records of colonies (1930 to 1995) at Sky & Isaac Lakes, Albemarle Twp.

Grey: there are recent reports of nesting at McNab Lake, Keppel Twp., and a record (1993) of 11 nests at a former site in Bognor Marsh, Sydenham Twp., prior to the recent "draw-down."

IDENTIFICATION: length 23-28cm; wingspan 57-65cm. A distinctive, small, dark tern of freshwater marshes. Adult (breeding): head and body mostly black, wings and tail silvery-gray; bill black; legs blackish-red. Immature: back gray with barred and scalloped brown wing coverts; underparts white with dark patch on sides of breast.

Voice: call is harsh, metallic *krik* and higher *kyip*.

Food: aquatic insects, small fish, snails, tadpoles and frogs. Takes food from surface and rarely plunge-dives.

Habitat: primarily large, freshwater marshes, sloughs and wet meadows for breeding; marine habitats (non-breeding).

Breeding: usually breeds in small, loosely-knit colonies, and in Ontario, most often in cattail marshes, and occasionally in bogs and flooded areas with marsh vegetation. Most nests are placed on floating mats of dead vegetation, but some were on wood pieces and logs. Nests are most often depressions in the substrate and are variously lined with wet pieces of marsh vegetation, mud, peat, moss, and sticks.

Movements: winters on both coasts of Central and South America and south to Peru. Arrives in spring in early to mid-May.

10. Short-eared Owl **G5 SC S3S4B**

Asio flammeus

There has been a long-term decline which warrants continuing scrutiny. Loss of habitat in southern part of range is cited as main reason for the population drop.

There are 22 species of owls in North America belonging to two families, Tytonidae: Barn Owls, and Strigidae: Typical Owls. The Short-eared Owl belongs in the Strigidae and is a cosmopolitan, widely-distributed species, occurring on every continent except Australia. It breeds across central Canada north to the tundra, south to the southern US states in the west, and to the central states in the east. It often hunts during daylight, and its buoyant, erratic flight low over the ground may remind one of a giant moth.

Local Status: Its scarcity in Grey-Bruce reflects its general decline in southern Ontario. A lack of permanent grassland habitat may be the primary reason for this decline.

Nest Records: *Grey*: a nest with young (1993), St. Vincent Twp. This is the only nest record from our area, although fledged young were reported from North Bruce, Saugeen Twp., Bruce County (1988).

IDENTIFICATION: length 37-38cm; wingspan 96cm. A medium-sized owl. Adult: tawny-brown colour of upper parts is mottled with buff spots and bars; breast is boldly streaked and lighter belly has finer streaks; large round head with small, rarely-seen ear tufts; facial disc gray-white with black around eyes. In flight, a black wrist mark on the underwing is a prominent field mark. Immature: crown and rump dark brown, facial disc brown-black; underparts warm buff.

Voice: silent except in breeding season; male gives muffled *poo poo poo*; alarm calls are sharp barks *keee-ow*.

Food: small mammals and, less frequently, birds.

Habitat: open country: tundra, grassland, marshes.

Breeding: in Ontario breeds in abandoned grassy fields, heath bogs, airport fields, cattail marshes, and tundra. Nests are slight depressions on the ground or merely cups of dried weeds and flattened grasses. Linings sometimes included a few feathers. Average-sized clutch is 4 to 7; eggs white.

Movements: winters generally in the breeding range: some move south from Canada as far as Mexico and southern Florida.

11. Red-headed Woodpecker **G5 SC S3B VUL**

Melanerpes erythrocephalus

This species has shown a steady decline in recent decades due to habitat loss, nest site competition, and the increase in automobile traffic.

The worldwide woodpecker family Picidae includes 25 species which occur in North America. This number includes species of accidental occurrence and the Ivory-billed Woodpecker which probably has been extirpated from North America, and is approaching extinction. The Red-headed Woodpecker, a striking and unmistakable species, breeds from southeastern Canada south to the southern United States, and generally east of the Rocky Mountains.

Local Status: In Ontario this woodpecker is found mainly in the southern deciduous forest region, and its general decline in the province is reflected in its comparative rarity in Grey-Bruce.

Nest Records: *Bruce*: a nest record (1996) at Cameron Lake, St. Edmunds Twp.

Grey: a nest (1943) in St.Vincent Twp. near Meaford; two nests (1947, 1948) at the same nest site near Garryowen, Sydenham Twp.; and a nest (1991) near Thornbury, in Collingwood Twp.

IDENTIFICATION: length 23.5cm; wingspan 43cm.
Adults: head, neck, and throat are bright red, and contrast with blue-black back and white underparts; white inner wing patches and white rump are apparent in both flying and perched birds. Immature: head and back are brown; underparts are whitish and sometimes streaked.

Voice: a loud, wheezy *queerp*, also a low rattle *krrrrr*; drum is weak, slow, and in short bursts.

Food: omnivorous. Diet includes a wide variety of insects, spiders, earthworms, seeds, nuts, fruit, and sometimes eggs and nestlings of other birds. Hawks for insects and may forage on ground.

Habitat: open woodland which is more often deciduous than coniferous; also orchards and groves of trees in farm country.

Breeding: in Ontario breeds in and at the edge of open woodlands; beaver-flooded areas; fence rows and treed roadsides in agricultural land; in city parks, ravines, golf courses, and residential gardens. Nest cavities are usually in dead trees which are invariably deciduous, but hydro poles and hollow posts are also used. Cavities are sometimes reused in successive years.

Movements: winters usually in the southern two-thirds of breeding range. Arrives in spring in mid-May.

12. Loggerhead Shrike **G5 END S2B END**

Lanius ludovicianus

The shrinking of suitable nesting habitat is one reason suggested for the decline of this species in the eastern part of its North American range.

The Loggerhead Shrike is one of three species of shrike in North America. Shrikes belong to the family Laniidae, and are predatory songbirds with hooked bills, that prey on rodents, birds and insects. The Loggerhead Shrike breeds from central Canada in Alberta, east to southern Quebec, and south locally to Florida, the Gulf coast, parts of Washington, Oregon, California and into Mexico.

Local Status: In Ontario most of the small, known breeding populations are now confined to a few southeastern counties. Nest records indicate that a few pairs persist in Grey-Bruce.

Nest Records: *Bruce*: a nest record (1937) at Lucknow, Kinloss Twp.; a nest record (1972) from near Wiarton, Amabel Twp.; a nest record (1985) from Arran Lake, Arran Twp.; a nest record (1987) from Hepworth, Amabel Twp.; a nest (1993) at Tara, Arran Twp.; and two nests (2000) near Allenford, Arran Twp.

Grey: a nest (1930) at Redwing, Collingwood Twp.; three nests (1937, 1972, 1983) from the Meaford area, St. Vincent Twp.; two nests (1968, 1969) near Coffin Hill, Sydenham Twp.; and a nest (1996) near Epping, Euphrasia Twp.

IDENTIFICATION: length 23cm; wingspan 30.5cm. Adults: dark gray bird with paler gray, unmarked underparts; large black mask, black wings and tail, both with white markings conspicuous mostly in flight; bill usually entirely black. Immature: grayish, and lightly barred on underparts like adult Northern Shrike.
Voice: song is a chat-like series of repeated notes, as *zee-ert* or *surp-see*; calls are a harsh *jaaa* or *teen raad raad raad* etc.
Food: mostly large insects (grasshoppers, crickets); also mice and small birds; and occasionally molluscs, amphibians, reptiles, crustaceans, and fish.
Habitat: semi-open country with hunting perches like utility wires, fences, and shrubs.
Breeding: in Ontario breeds in agricultural areas in fields, pastures, and old orchards; also in fencerows overgrown with shrubs and small trees. Hawthorns are the preferred nest shrub, and, more recently, nests are often in Red Cedar, *Juniperus virginiana*, in southeastern counties. Nests are woven, bulky cups of sticks, twigs and grasses, and are lined with feathers, hair, grasses, plant down and rootlets. Average nest heights were 1.7 to 3m.
Movements: winters south to the southern limits of the breeding range. Arrives back in spring from late March to mid-April.

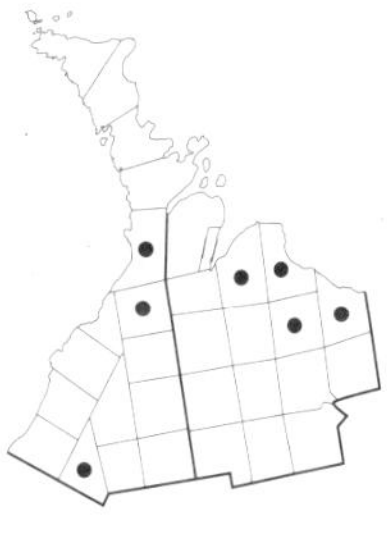

13. Prairie Warbler G5 NAR S3S4B

Dendroica discolor

Designated Special Concern Category in 1985, it was delisted in 1999 because there had been little change in its numbers and habitat availability in its most important breeding area in Ontario, namely on the eastern shore of Georgian Bay. In that area its preferred nesting habitat is scrub oak-juniper.

The New World wood warblers are mostly colourful tropical birds which migrate north to our temperate latitudes in order to breed. After breeding they migrate back to their southern homes. In North America 57 species in the family Parulidae occur, including accidentals and the presumed extinct Bachman's Warbler. The Prairie Warbler breeds in eastern North America from southern Ontario, northern New York and southern Maine south to Florida and the Gulf states.

Local Status: In Ontario the Prairie Warbler is at the northern limit of its breeding range, and its specific habitat requirements further limit its breeding population in the province. It probably no longer breeds in Grey-Bruce.

Nest Records: *Bruce*: a nest (1953) on Lot 13, Conc. I, St. Edmunds Twp. This is the only nest record of this species for our area.

IDENTIFICATION: length 12cm; wingspan 18cm. Tail-bobbing is a good identification feature. Male: back yellowish olive-green, underparts bright yellow with black streaks on sides and two black facial marks through and below eye; there are two yellow wing bars and outer tail feathers are white. Female; similar, but duller with less distinct dark streaking. Immature: duller still; streaking faint, and cheek often gray.

Voice: song is a thin, ascending *zee zee zee zee zee zee.*

Food: mostly insects of various kinds; nestlings are fed mainly caterpillars. They also eat a few berries, and occasionally sap from sapsucker drillings.

Habitat: brushy slashings, bushy pastures; low pine barrens, scrub oak-juniper.

Breeding: in Ontario, usually in areas of open low growth, with juniper and scrub oak predominant, and less often with shrubs such as hazel, dogwood, and viburnum. These areas are sandy or rocky, and both are usually dry. Nests were in low shrubs at an average height of 0.6 to 1.0 m. Nests are cups of grasses, bark strips, and plant fibres, and were lined with hair, plant down, rootlets and fine grasses.

Movements: winters from Florida and the Gulf coast south to the West Indies and Central America. Arrives in spring in early to mid-May.

14. Cerulean Warbler G4 SC S3B VUL

Dendroica cerulea

One of the large group of colourful wood warblers, the sky-blue plumage of the Cerulean Warbler is difficult to observe during the breeding season, since it stays in the high canopy of large deciduous trees where it nests and forages. It breeds from southeastern Ontario and the northeastern United States south to North Carolina and northern Alabama, Mississippi and southern Arkansas.

Local Status: A Carolinian species, toward the northern limit of its range in Ontario, its declining populations are due, in part, to decimation of large tracts of mature, deciduous woodlands, where it breeds.

Nest Records: *Bruce*: a nest (1953), 16 km W of Walkerton, in Greenock Twp. This is the only nest record reported for Grey-Bruce (Moore, 1954).

IDENTIFICATION: length 12cm; wingspan 20 cm. Blue or blue-green plumage is distinctive. Male: light bluish above, underparts white with side streaks and a dark band across chest; white wing bars. Female: dull bluish-green upper parts, dull whitish underparts; white eyebrow stripe. Immature: similar to females, but more greenish above and yellowish below.
Voice: song a high-pitched, ascending buzz *tzeedl tzeedl tzeedl ti ti ti tzeeeee.*
Food: mostly insects.
Habitat: mature, deciduous forests.
Breeding: in Ontario breeds in second growth and mature deciduous woods. The preferred nest trees were maple, oak, elm, and basswood. Nests were usually high up, with average heights of 8.8 to 13.1m; one nest was at a height of 20 m. Nests were neat, shallow cups saddled on horizontal branches, out from the main trunk.
Movements: winters from northern South America south to Peru. Average arrival date in spring is mid-May.

SUPPLEMENTARY LIST

Ruddy Duck G5 S2B

A western species with a small breeding population in Ontario, the Ruddy Duck has been recorded breeding at Dundalk, Proton Twp., Grey County in 1989 and 1990.

Virginia Rail G5 S4B, Common Moorhen G5 S4B, American Coot G5 S4B

These members of the family Rallidae are marsh-dwelling species and their breeding populations and nests are uncommon in Grey-Bruce, due in part to the relatively small amount of suitable habitat.

Sandhill Crane G5 S4B

Although its Ontario breeding populations appear to be increasing, the Sandhill Crane is still a scattered and uncommon breeding bird in southern Ontario. Breeding observations in Bruce County have been recorded for Kinloss/Greenock Twps., and Lindsay Twp. (1985), although no nests have been found.

Wilson's Phalarope G5 S3B

This western shorebird has received a recommended rare status in Ontario in the Ontario Birds At Risk listing, and has a small number of known breeding locations in the province. There is a report of breeding near Wiarton in Amabel Twp., Bruce County in 1981, and two young with an adult were seen at Isaac Lake, Albemarle Twp., Bruce County on 6 July, 1977.

Great Black-backed Gull G5 S2B

This marine gull has bred since 1954 in Ontario at a few locations in the Great Lakes region. All breeding and nesting records from Grey-Bruce have occurred in Bruce County on islands in Lake Huron, from 1954 to 1999.

Yellow-bellied Flycatcher G5 S5B

Although this ground-nesting, bog-dwelling flycatcher of the boreal forest region is not presently at risk, it is included here because its most southerly provincial nesting in modern times occurred in 1982 in Sullivan Twp., Grey County.

Northern Parula G5 S4B

This warbler has no recommended risk category, but it is an uncommon breeder in Ontario, primarily north of the southern limit of the Canadian Shield. In Grey-Bruce there is only a single record in 1966 of an unsuccessful nesting attempt at Red Bay, Albemarle Twp., Bruce County.

Cape May Warbler G5 S5B

Another northern warbler with no recommended risk category. Only five nests of this species have been found in Ontario. The Grey-Bruce area has the distinction of having the first Ontario nest record and one of the first North American nests. This nest was found in 1934 at Dorcas Bay, St. Edmunds Twp., Bruce County, by Dr. Paul Harrington of Toronto (Bent, 1953).

Literature Cited

American Ornithologists' Union. 1998. **Check-list of North American Birds. 7th ed.** American Ornithologists Union.

Austen, M.J.W., M.D. Cadman, R.D. James 1994. **Ontario Birds at Risk**, Federation of Ontario Naturalists and Long Point Bird Observatory.

Bent, A.C.1953. **Life histories of North American Wood Warblers.** Bulletin 203, US Nat. Mus.

Committee on the Status of Endangered Wildlife in Canada. **Canadian Species at Risk, April 2000.** COSEWIC 2000.

Kirk, M.D. and G.K. Peck. 1984. **Botanical and Ornithological Studies: Interim Report. Northern Fishing Islands and Oliphant Fens.** North Grey Region and Sauble Valley Conservation Authorities.

Mitchell, M.H. 1935. **The Passenger Pigeon in Ontario.** Contribution of the Royal Ontario Museum of Zoology 7: 1-181.

Moore, G.A. 1954. **Cerulean Warbler and Blue-gray Gnatcatcher Nesting in Bruce County, Ontario.** Canadian Field-Naturalist 68: 44.

Peck, G.K. and R.D. James.

1983. **Breeding birds of Ontario: nidiology and distribution.** Volume 1: Nonpasserines. Life Sciences Miscellaneous Publications. Toronto, Royal Ontario Museum. 321pp.

1987 **Breeding birds of Ontario: nidiology and distribution**. Volume 2: Passerines. Life Sciences Misc. Publications. Toronto, Royal Ontario Museum. 387 pp.

1993a **Breeding birds of Ontario: nidiology and distribution**. Volume 1: Nonpasserines (First revision — part A: loons to ducks). Ontario Birds 11: 18-22.

1993b **Breeding birds of Ontario: nidiology and distribution.** Volume 1: Nonpasserines (First revision — part B: vultures to phalaropes). Ontario Birds 11: 83-91.

1994 **Breeding birds of Ontario: nidiology and distribution**. Volume 1: Nonpasserines (First revision — part C: jaegers to woodpeckers). Ontario Birds 12: 11-18.

1997 **Breeding birds of Ontario: nidiology and distribution.** Volume 2: Passerines (First revision — part A: flycatchers to gnatcatchers). Ontario Birds 15: 95-107.

1998a **Breeding birds of Ontario: nidiology and distribution.** Volume 2: Passerines (First revision — part B: thrushes to warblers). Ontario Birds 16: 11-25.

1998b **Breeding birds of Ontario: nidiology and distribution.** Volume 2: Passerines (First revision — part C: tanagers to Old World sparrows). Ontario Birds 16: 111-127.

1999 **Breeding birds of Ontario: nidiology and distribution.** Volume 1: Nonpasserines (Additions and Revisions) Appendix C. Ontario Birds 17: 105-123.

Williams, M.Y. 1942. **Notes on the fauna of Bruce Peninsula, Manitoulin and adjacent islands.** Canadian Field-Naturalist 56: 70-81.

Selected Bibliography

Baicich, P.J., and C.J.O. Harrison. 1997. **A Guide to the Nests, Eggs, and Nestlings of North American Birds. 2nd ed.** Academic Press, San Diego.

Cadman, M.D., P.F.J. Eagles, and F.M. Helleiner (compilers). 1987. **Atlas of the Breeding Birds of Ontario.** University of Waterloo Press, Waterloo, Ontario.

Del Hoyo, J., A. Elliott and J. Sargatal (eds.). 1992-99. **Handbook of the Birds of the World.** Lynx Edicions, Barcelona.

Farrand, J. Jr. 1983. **The Audubon Society Master Guide to Birding (3 volumes).** A. A. Knopf, New York.

Kaufman, K. 1996. **Lives of North American Birds**. Houghton Mifflin, New York.

Godfrey, W.E. 1986. **The Birds of Canada, rev. ed.** National Museums of Canada, Ottawa.

McNicholl, M.K., and J.L. Cranmer-Byng (eds.). 1994. **Ornithology in Ontario. Special Publication No. 1.** Ontario Field Ornithologists.

Sibley, D.A. 2000. **The Sibley Guide to Birds**. National Audubon Society. Alfred A. Knopf, New York.

Note: Many excellent sound recordings are commercially available which will aid in voice identification. Similarly, there are a number of other bird field guides in bookshops or public libraries which will provide colour pictures, of North American birds.

MAMMALS

With as many as 46 different mammal species, Grey and Bruce counties are home to more than half of the province's mammals. Part of this wealth of species is due to the abundance of larger natural areas that remain in the counties, and part of it is simply due to geography. Grey and Bruce Counties are situated at the dividing line between what typically are 'northern' species and 'southern' species, meaning that this area is at the southern or northern edge of the range for many of these species. Mammals such as the Fisher and River Otter are at the southernmost extent of their Ontario range here, whereas the Eastern Cottontail, Gray Squirrel, European Hare and Virginia Opossum are near or at the northern extent of their range.

Although not all of the 46 species recorded in Grey and Bruce are common in the area, only one of them is considered rare at the provincial level based on its NHIC 'S' ranking; that species is the Eastern Small-footed Bat (*Myotis leibii*). One other species, the Northern Long-eared Bat (*Myotis septentrionalis*) is listed as 'S3?' indicating that its status is uncertain. Based on recent surveys throughout the province, however, the Northern Long-eared Bat will likely be placed in the 'S4' category.

The fact that only one mammal in Grey and Bruce Counties can be considered provincially rare, does not mean that we should be lulled into a sense of security about the status of our mammals. As pointed out earlier, Grey and Bruce are at the extreme northern or southern edge of the range of many species, and any change in the amount or quality of habitat could result in a loss of one or more species from the area. In particular, the Bruce Peninsula is home to species that are otherwise restricted to the shield and its associated large tracts of contiguous natural habitat (Fisher and River Otter). Fragmentation of the Peninsula and the associated encroachment of development has the potential to make the area less suitable for these species. The preservation of the quality and quantity of habitat in Grey and Bruce are the key elements to maintaining this high level of biodiversity.

Selected References

Banfield, A.W.F. 1974. **The Mammals of Canada**. University of Toronto Press, Toronto.

Dobbyn, J.S. 1994. **Atlas of the Mammals of Ontario**. Federation of Ontario Naturalists, Don Mills.

Forsyth, A. 1985. **Mammals of the Canadian Wild**. Camden Publishing Ltd., Camden East.

Kurta, A. 1995. **Mammals of the Great Lakes Region**. Fitzhenry and Whiteside Limited, Markham.

Peterson, R.L. 1966. **The Mammals of Eastern Canada**. Oxford University Press, Toronto.

Van Zyll de Jong, C.G. 1985. **Handbook of Canadian Mammals. 2. Bats**. National Museum of Natural Sciences, Ottawa.

BATS

Grey and Bruce Counties are rich in bat species, having seven of the eight resident Ontario species. The only one not living here is the Eastern Pipistrelle, which is most commonly found in southern and eastern Ontario.

The amount of undisturbed habitat does not seem to be as crucial for bats as it is for some other mammal species. Some bats do as well in urban areas as they do in extensive natural areas; some possibly better. An abundance of roosting sites, clean water and a good supply of appropriate insects seems to be sufficient for most species. What may be more important to bats is the availability of suitable hibernation sites (hibernacula). While three out of the eight Ontario bats migrate south each winter, the remaining five stay in the province and hibernate, seeking out caves, crevices or abandoned mines as hibernacula. There are no mine sites here that are suitable for bats; but the old water purification filter bed near Inglis Falls has been used to an extent by Big Brown Bats (*Eptesicus fuscus*).

Spelunking or caving is becoming more and more popular, and cavers are often the first to discover previously unknown hibernacula, thereby expanding our knowledge of the distribution of bats. Spelunking can, however, impact on the survival of wintering bats if cavers do not keep the best interests of the bats in mind. Ethical cavers will not enter a cave in winter if that cave is known to support hibernating bats, and will immediately leave if they discover bats in a cave where they had not previously been found.

Bats are true hibernators, meaning that they drastically lower their metabolism including their heart rate, respiration and body temperature. Preferred hibernation sites are those caves or mines where the ambient temperature does not drop below zero. This means that bats will not have to expend any energy to maintain a body temperature above freezing. Even short visits to a cave will result in some of the bats awakening, resulting in increased metabolism and energy expenditure. This increased energy output can be quite significant, even for the short time that they remain awake. Repeated visits to the cave and subsequent awakenings, may mean that a bat will consume all of its available fat reserves prior to spring when there is an adequate number of insects to feed on, ultimately resulting in starvation.

Bats are a difficult group of mammals to census. They are nocturnal, hard to identify and not always easy to find. This means that the status of many bats may be based on an artificially limited number of records, resulting in an assigned status that is actually lower than reality. As a group, bats are in need of a great deal more study to obtain better assessments of their status, and to determine their conservation needs.

1. Eastern Small-footed Bat G3 S2S3

Myotis leibii **Family Vespertilionidae**

[Myotis = Mouse Eared; leibii — after Dr G.C.Leib]

The Eastern Small-footed Bat is the smallest bat in Ontario, weighing in at a mere 4 g — approximately equal to the weight of a quarter. With its proportionately small hind feet as compared to other Ontario species, it is also appropriately named. The first record of the Eastern Small-footed Bat in Ontario was not made until 1929 when a single individual was discovered in a shed near London Ontario. Since that time, several hundred individuals have been found, but primarily in a few hibernacula in eastern Ontario. As such, the species could be considered rare to very rare in the province. Records obtained for the *Atlas of the Mammals of Ontario* show the Ontario range for this species as being generally restricted to the southern part of the province, south of the north shore of Georgian Bay. Outside of Ontario it is limited to the northeastern United States, and a small section of southwestern Quebec.

The Eastern Small-footed Bat tends to hibernate in cooler, drier sites than the Little Brown Bat and tends to enter hibernation later (late November) and emerge earlier (April). It has also been known to hibernate under rocks on the cave floor, in low crevices and under overhangs — areas not normally utilized by other bats. These habits may result in the Eastern Small-footed Bat being overlooked or missed during winter cave surveys, thereby exaggerating its rarity. Eastern Small-footed Bats have been trapped in both Rovers and Stream Cave on the northern Bruce Peninsula during fall swarming, but not found in these caves during the winter. A similar pattern has been observed at the few abandoned mines north of Georgian Bay where the species has also been found during fall swarming. This may be an indication that these bats migrate further south prior to hibernation. Other data suggests that Eastern Small-footed Bats may not travel far from their hibernation sites. This may restrict the species to areas with adequate caves, mines or other suitable hibernacula. From this standpoint Grey and Bruce Counties are prime areas for this species due to the abundance of karst topography in the region and the resulting abundance of suitable hibernation habitat.

Local Distribution: Uncommon in both Bruce and Grey Counties, with six known records from Bruce and seven from Grey. The records from Bruce County were of bats swarming into Rovers and Stream Cave on the Bruce Peninsula. The Grey County records were of foraging bats in July from various locations in the county ranging from Inglis Falls south to Durham. This bat is likely more common than would appear due to the difficulty in identifying bats in general, and the secretive nature and non-typical hibernation habits of the species.

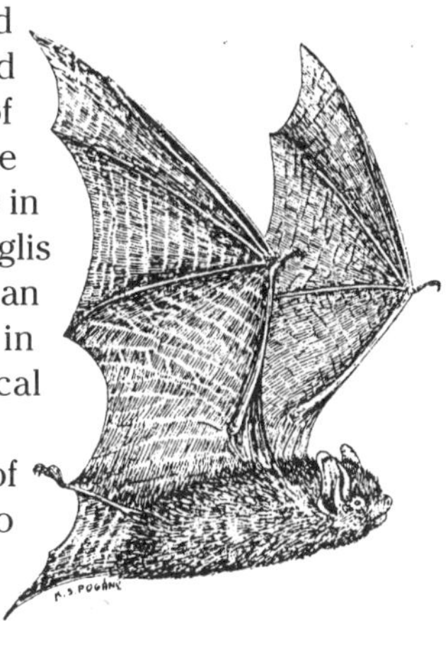

Habitat: Little is known about the summer habitat of this bat, but it is likely found in mixed forests adjacent to water and within a reasonable distance of suitable hibernation sites. Roosting may occur under rock ledges, in hollow trees or occasionally in buildings. Hibernation is in dry, cool caves, rock crevices or other protected sites where the temperature remains above -9 deg. C.

DESCRIPTION
Weight: 4 g.
Size: Wingspan: 21-25 cm; Forearm: 31-34 mm; Hind Foot: 6-9 mm.
Fur silky, glossy yellowish or golden brown with a black 'mask' around the face. Ears and flight membranes black. Identification based on a forearm length of <34mm and hindfoot < 9 mm.
Hibernation: November to April.

PLANTS

An unusually large number of rare plants are listed for this area, especially on the Bruce Peninsula. To a large extent this is due to the variety of habitat types, but also to the configuration of the Great Lakes, and to the north-south corridor created by the Niagara Escarpment. Very soon after this area became colonized by Europeans its unusual qualities and the diversity of its flora was recognized by botanists. The endemic species found along the Lake Huron shore attracted attention and rarities such as the Hart's Tongue Fern drew botanists from distant places. For the less technically minded the wonderful displays of orchids in late spring and early summer brought sightseers. Unfortunately, these flowers and species such as Ginseng, were also attractive to plant hunters out to make a profit. Luckily, distance and the difficulties of transportation in earlier days prevented the total extirpation of the targeted species. The development of shoreline cottages has destroyed much valuable habitat but, again, changing attitudes in the latter half of the last century recognized the need to protect as much of this unique area as possible so that we now have the National Park, provincial parks, conservation areas and nature reserves. It is hoped that the information on the plants listed here will help in maintaining this national treasure.

It should be noted that, due to the rarity of these plants, it has not been possible in many cases to check their dimensions in the field. The numbers quoted in the descriptions are gleaned mostly from standard works of reference (see references) and may differ slightly from what is actually encountered. Usually, however, the proportions will remain the same, i.e. length to width. Similarly, it is difficult to be sure of the actual flowering times for some species.
It has also been difficult to obtain illustrations so, in a number of cases, photographs of herbarium specimens have been used. Many of these species are not described in general floras covering eastern North America, so it is hoped that the material presented here, however imperfect, will be helpful in identification.

Under the heading **Local Distribution** the methods used to determine the degree of rarity (from Common to Extremely Rare) are complicated, but they are based on long experience in the field by naturalists (mainly professional botanists), and will serve as a basis for comparison between species. The terms refer specifically to the degree of rarity in Grey and Bruce Counties. As noted before, some globally or provincially rare species are not uncommon in this area.

FERNS

Bruce and Grey Counties with 50 species, have an exceptional number of kinds of fern. Among these are seven species which are considered to be rare in Ontario and, in some cases, in North America, yet some of them are not too difficult to find. Bear in mind, when dealing with ferns, that they do not have flowers, so the shape of the frond (leaf) becomes a very important tool for identification. Most fern fronds are divided into leaflets. If they are simple leaflets, the fern is said to be "once divided," if the leaflets are divided into subleaflets, it is "twice divided," and if the sub-leaflets are divided again the fern is "thrice divided." Ferns reproduce by spores. These develop in sporangia, which are borne in structures which are different for each species, so this is another important feature to look for. Complete descriptions of all local species are to be found in **The Ferns of Grey and Bruce Counties**.

Ferns are divided into families by botanists. Four of them are represented here, the Adder's Tongue (Moonwort, Grape Fern) family **Ophioglossaceae**, the Maidenhair Fern family **Pteridaceae**, the Spleenwort family **Aspleniaceae**, and the Wood Fern family **Dryopteridaceae**.

Useful References:

Bruce-Grey Plant Committee. 1999. **A Guide to the Ferns of Grey and Bruce Counties, Ontario.** Bruce-Grey Plant Committee, Owen Sound Field Naturalists.

Cobb, B. 1963. **A Field Guide to the Ferns**. Peterson Field Guide Series. Houghton Mifflin Co. Boston.

Cody, W.J. & Britton, D.M. 1989. **The Ferns and Fern Allies of Canada**. Agriculture Canada.

Broad Beech Fern **G5 SC S3**

(Phegopteris hexagonoptera)

A specimen was collected in 1993 in the Lion's Head area, and is in the herbarium (TRT) at the Royal Ontario Museum in Toronto. A description can be found in **A Field Guide to the Ferns** (Cobb, 1963). The location was not clearly pinpointed, and a search in 1999 could not relocate it.

1. Blunt Lobed Grape Fern **G4? S3**

***Botrychium oneidense* (Gilbert) House** **Adder's Tongue Family**

Since it was first separated from other Grape Ferns about seventy-five years ago, the Blunt Lobed Grape Fern has been a continual source of taxonomic disagreement. Many botanists have wished to interpret it simply as a variety of Leathery Grape Fern (*Botrychium multifidum)* or Dissected Grape Fern (*B. dissectum*). In a study some years ago, samples from over 60 localities in Ohio, Ontario and Michigan were taken. Over 4500 specimens of *B. dissectum* and 2000 of *B. oneidense* were compared for habitat, range, leaf blade, colouring, and root. Differences were found in all categories. Blunt Lobed Grape Fern has paler and more delicate roots; duller, smoother leaves with fewer large, rounded segments, with nearly entire to shallowly-cut margins. Dissected Grape Fern, has thicker, darker roots; shiny leaf blades deeply cut into numerous sharply pointed segments. The spores of Blunt Lobed Grape Fern mature in October, those of Dissected Grape Fern in September. From these results, the opinion of experts is that it should be considered a different species, so we are describing it as such here.

This eastern North American fern is on the western edge of its range here. It is found in low, damp shady woods, also in swamps with acid soil. It is not found in open fields. It seems to be rare in Grey and Bruce Counties as well as in Ontario generally. This may be due partly to its recent acceptance as a separate species, as well as the difficulty of separation from Dissected Grape Fern and Leathery Grape Fern, with both of which it is frequently found growing. However, this mixing of species is not uncommon among other Grape Ferns.

DESCRIPTION

Height: ± 25 cm including fertile segment.

Growth Form: One frond branching just above the ground.

Sterile Portion: Blade dull bluish-green. Somewhat leathery. 3 divisions, each divided into 1-3 pairs of broad ovate subdivisions, slightly lobed at the apex. Margins entire, wavy, or shallowly toothed. Veins not showing clearly.

Fertile Portion: Long stem, 2-3 x length of blade. Once or twice divided. Sporangia borne in 2 rows on reduced leaflets.

Season: Frond overwinters.

Mostly remains green (related species turn bronze). Spores mature in October.

2. Purple Stemmed Cliffbrake G5 S3

***Pellaea atropurpurea* (L.) Link Maidenhair Fern Family — Pteridaceae**

[*pelos* — dark (Gr.) — referring to grey-green leaves; *atro* — dark, *purpureus* — red-violet (Lat.) referring to the stem]

Both the English and Latin names of this species centre on the dark purplish stem. The only species with which it could be confused in this area is Smooth Cliffbrake, which is much commoner, and also has dark, shiny stems. A better clue is the bristly hairiness of the stalks of Purple Stemmed Cliffbrake. This can be felt by running the fingers down the stem. Smooth Cliffbrake has shiny stems that feel smooth. In addition, Purple Stemmed Cliffbrake is an example of a fern that produces two different looking kinds of fronds — one with spores and the others smaller and sterile. Fronds with spores in Smooth Cliffbrake are identical with the sterile ones. Purple Stemmed Cliffbrake is widely distributed from New England across the eastern and southern United States, extending down into Mexico and central America. In Canada, it is only found in Quebec and Ontario. Its occurrence is sporadic but it is found no further north than Manitoulin Island, Nipissing District and Renfrew County. It is also in Frontenac, Leeds and Niagara, all of which have calcareous rocks with suitable habitat for this plant. Thus, it is very much on the northern edge of its range in our area.

Local Distribution: Rare. It has only been found in the four northern township units on the Bruce Peninsula.

Local Habitat: Dry crevices in rock, with little or no soil. On dolostone outcrops, alvars, or occasionally at or below the escarpment cliffs. (Smooth Cliffbrake mostly grows on cliffs).

DESCRIPTION

Growth Form: Erect fronds with purplish-brown bristly stalks arising stiffly from an erect, compact rhizome.

Sterile Frond: Blade bluish-green, leathery. Partly twice divided.
Length ± 15 cm, width ± 3 cm.

Leaflets: Ovate-oblong. Lower ones may be divided into subleaflets.

Fertile Frond: Dark blue-green, leathery. Partly twice divided. Length ± 20 cm, width ± 4 cm.

Leaflets: Linear-oblong. Edges rolled under.

Sporangia: Concealed in rolled-under edge of subleaflet.

Season: Evergreen.

3. Wall Rue

G5 S2

***Asplenium ruta-muraria* L.**

Spleenwort Family — Aspleniaceae

[*ruta* — rue (herb), *muraria* — wall (Lat.)]

This tiny, delicate, deep green fern with its fan-shaped leaves is our smallest fern and the Bruce Peninsula is one of the few places in Canada where it occurs. It is mainly an amphi-Atlantic species being found in New England and the Appalachians, but extending as far west as Illinois and Michigan in suitable habitats. It is very much commoner in Europe — especially in the British Isles, but it extends east into Italy and Germany. The prevalence of old stone walls in European countries favours it. The English name is derived from its exploitation of this artificial habitat. The name *Asplenium* was coined nearly 2000 years ago by the Greek physician Dioscorides. Plants in this family were supposed to cure diseases of the spleen, hence Spleenworts. Wall Rue was once used to treat scrofula (tuberculosis of the neck glands).

Local Distribution: Rare. Known on the Bruce Peninsula from Hope Bay north, at about 15 locations, almost all of which are on the Niagara Escarpment. Most easily seen in the Bruce Peninsula National Park, including Flowerpot Island.

Local Habitat: Tiny cracks or seams in dolostone. Often on the sides of large boulders or shallow rock ledges.

DESCRIPTION

Growth Form: Fronds arising in a cluster among persistent stalk bases, from a tiny rhizome embedded in the rock.

Fronds: ± 6 cm long, ± 3 cm wide. Stalk reddish-brown, dark brown scales at base, shading to green. Blade leathery, green, thrice divided.

Leaflets: Main leaflets stalked, divided into 2 or 3 wedge-shaped sub-leaflets, which are toothed.

Sporangia: In elongate strips on back of sub-leaflets.

Season: Evergreen.

4. American Hart's Tongue Fern G4 SC S3

***Asplenium scolopendrium* L. var. *americanum* (Fern.) Kart.& Ghandi** **Spleenwort Family**

[*Phyllitis scolopendrium* (L.) Newman var, *americanum* Fernald]

Hart's Tongue Fern is much easier to find in our area than Wall Rue, but it is not widespread in North America. It is found in significant quantities only in Grey, Bruce and Simcoe Counties, although it also occurs further south along the escarpment in Ontario, and there are a few small, isolated, populations in four of the eastern states as far apart as Michigan and Alabama. Reports from other localities seem to be introductions of European Hart's Tongue which persist temporarily. American Hart's Tongue Fern is a very good example of a globally rare plant which appears to be comparatively common here. Like Wall Rue, this whole species has an amphi-Atlantic distribution, although the North American variety is slightly different from the Hart's Tongue found on walls and rocks in the British Isles, and much less commonly in other western European countries. It exhibits an affinity for the Niagara Escarpment, although it does occur away from it in a number of places. It is probably the combination of dolostone rock outcrops and a moist climate, including a heavy protective snowfall, which make it so well established here. Additionally, the lack of development compared with other localities has enabled it to persist. This is one of the few ferns in which the frond is not divided at all. The large tongue-shaped fronds are easy to pick out in the shady forest.

Local Distribution: Very uncommon overall in Bruce County. Grey County, on the other hand, probably has much more than half the total population of the New World. It can be found in suitable habitat throughout much of the county. In Bruce County, it occurs mainly on the Peninsula, but in the southernmost part of the county it is reported near the Teeswater River and at two sites near Greenock Creek, much farther from the escarpment than anywhere else in Ontario.

Local Habitat: Rich upland deciduous forest, usually dominated by Sugar Maple, in deep shade. On both dolostone bedrock and fragments (including boulders) transported locally by glaciation. On the Peninsula south to Owen Sound it is usually on bedrock. Further southeast in Grey County more than half the populations are on dolostone rubble, sometimes in rocky ravines.

DESCRIPTION

Growth Form: Fronds arising in a cluster from a short surface rhizome, anchored by a mass of wiry roots.

Fronds: Length ± 40 cm, width ± 5 cm. Stalk brown, hairy. Blade oblong, rounded point, ear-shaped base. Dark green, glossy, leathery.

Sporangia: In elongate strips along the veins, forming a chevron pattern on underside of frond.

Season: Evergreen.

5. Robert's or Limestone Oak Fern G5 S2

Gymnocarpium robertianum **(Hoffm.) Newm.**
Wood Fern Family — Dryopteridaceae
[*gymnos* — naked, *karpos* — fruit (Gr.)]

Robert's Fern is a fern which has a much commoner relative — the Oak Fern. They both have a "triune" frond, in which the bottom pair of leaflets is as big as the rest of the frond. This makes them fairly easy to separate from all other ferns. They look a little like a miniature bracken, but are much smaller and more delicate. In our area, the Oak Fern is very much an inhabitant of moist forest floors, whereas Robert's Fern tends to grow in partly shaded rock crevices, often under conifers. This species is found from Newfoundland to the Great Lakes states. The Bruce Peninsula is its southern limit in Canada.

Local Distribution: Rare. Robert's Fern has only been found at about 12 locations in the two northernmost townships of the Bruce Peninsula.

Local Habitat: Dolostone, mostly outcrops and pavements, usually in moderate to light shade of mixed to coniferous woods. Once started it will often spread and fill a crevice.

DESCRIPTION

Growth From: Fronds arising singly from a slender rhizome.

Fronds: Length ± 25 cm, width ± 15 cm. Blade outline triangular. Wiry stalk and blade glandular. Thrice divided.

Leaflets: Triangular, longer than wide. Lower 2 pairs short-stalked. Upper pairs more elongate, very short stalked.

Subleaflets: Lanceolate, pairs subequal, lower one larger. Subsessile. Deeply divided.

Sporangia: In tiny round groups with no indusium (covering), on underside of frond.

Season: Fiddleheads appear in spring. Fronds die off in early fall.

6. Laurentian Fragile Fern G3G4 S2S3

***Cystopteris laurentiana* (Weatherby) Blasdell**
Wood Fern Family — Dryopteridaceae
[*cystos* bladder, *pteris* — fern (Gr.)]

There are four species of this group in our area. The commonest is Bulblet Fern, which is a long-bladed, lacy fern often seen hanging down from moist rock faces, or on rocky forest floors. It produces "bulblets" on the back of the frond, from which it can reproduce asexually. Fragile Fern is much smaller, though also lacy. It never produces bulblets, and is only found growing on rock. Laurentian Bladder Fern originated as a cross between Bulblet Fern and Fragile Fern, so it is between the two in size. It does occasionally produce deformed bulblets. Like Bulblet fern it has glandular hairs on the stalk and blade, which can be felt or seen with a lens. The fronds tend to round off at the bottom, unlike Bulblet Fern which is almost perfectly triangular. This species was first thought to be exclusively in the area of the Great Lakes; however it is now known from the Driftless Area of the Mid-west and in New England, as well as the Maritimes and Quebec, but it is not common anywhere. It can be confused with large specimens of Fragile Fern or with Mackay's Fragile Fern.

Local Distribution: Rare. Reported from all township units on the Bruce Peninsula except for Amabel and Sarawak, and from St. Vincent Township in Grey County. Also on the islands between the Bruce Peninsula and Manitoulin Island, (as well as on Manitoulin Island).

Local Habitat: Always on limestone or dolostone, moist cracks and ledges.

DESCRIPTION

Growth Form: Fronds clustered at the top of a short, scaly rhizome.

Fronds: Length ± 40 cm, width ± 12 cm. Dark base of stalk shading to reddish. Glandular hairs on blade and stalk. Thrice divided — lacy. Outline ovate — sharply tapered.

Leaflets: Lanceolate outline, stalked, opposite.

Subleaflets: Ovate outline, shallow rounded lobes.

Sporangia: In 1 mm wide dots on back of subleaflets. Larger than the parent species (Bulblet and Fragile).

Season: Slim fiddleheads appear from dead stalks in spring. Fronds die at first frost.

7. Oregon Woodsia

G5 S3

Woodsia oregana* D.C.Eaton ssp. *cathcartiana
(B.L.Robinson) Windham

Wood Fern Family

Oregon Woodsia is a very rare fern in our area. As its name suggests, it is found mainly in the west. Together with the other subspecies *oregana*, they occur throughout the western states and into B.C. and the Canadian Prairies, but only subspecies *cathcartiana* extends into Ontario and Quebec. At some point in time, the chromosomes of the original species have doubled and this is now a permanent feature of our subspecies. Discernible differences are larger spores and distinctly glandular stems and leaves. Manitoulin Island and the Bruce Peninsula are two of the few areas where it is found in eastern North America. There is also one station in Quebec and several along the Ottawa River. It is arguable whether this is the last remnant of a once more extensive distribution, or whether this is the first evidence of a new species moving in. The former seems most likely, as no new stations have been reported here for many years. Its more acid-tolerant relation *Woodsia ilvensis* or Rusty Woodsia is much more widespread but it is not found in this area. As its name implies, Rusty Woodsia is easily distinguished by its rusty hairs and scales, and it is not glandular. It is characteristic of *Woodsia* species that the stalk bases remain when the leaves die and break off in the autumn. In *Woodsia ilvensis* these stalks are equal in length — often referred to as an "even stubble." In *Woodsia oregana* ssp. *cathcartiana* the stalks are unequal i.e. an "uneven stubble."

Local Distribution: Extremely rare. Known historically from the Cabot Head area on the Bruce Peninsula, also currently from the Jones Falls area of Derby Township in Grey County.

Local Habitat: Calcareous rocks. In crevices on the Niagara Escarpment.

DESCRIPTION

Growth Form: Fronds in a small cluster among the uneven stubble of old leaf bases.

Fronds: Length ± 13 cm, width ± 3 cm. Stalk reddish-brown base, straw colour above, glandular, rarely with scales. Outline narrowly ovate. Thrice divided.

Leaflets: Glandular, ovate-triangular. Short stalked. Opposite.

Subleaflets: Oblong with rounded tip.
3-4 pairs. Larger ones with rounded lobes.

Sporangia: In round groups with a covering of elongate threads.

Season: Fronds die back early in fall, leaving uneven leaf bases. New fronds emerge in spring.

FLOWERING PLANTS — DICOTYLEDONS

This major group of flowering plants is characterized by having two "seed leaves" or cotyledons on germination. There are many hundreds of families within this group, all with distinct separating characteristics. They can be broadly divided into two groups: those families with "regular" flowers i.e. radially symmetric, which can be divided into two in any plane and will give two equal halves, and those families with irregular flowers which are bilaterally symmetric and can only be equally divided in one plane. The first nine species listed are in the Aster Family, details of which are given here. Number 17 to number 30 have regular flowers, and number 31 to 35 irregular flowers.

ASTER FAMILY — ASTERACEAE

This very large (10 000 species), highly evolved family is found throughout the world. Describing the species occupies about one-eighth of the average North American flora. Partly because it has penetrated so many different niches, some of its species are not only rare but their very existence is threatened. Additionally, some of the species we have listed here are on the edge of a much wider distribution, making them rare in this area. In order to recognize members of this family, it has to be clearly understood that the *flower* that you see in, for example, an Aster or a Daisy, is not simply one flower; it is a whole bunch of tiny flowers. In this book we refer to them as "Flower Heads"; the correct term is "capitulum." The tiny flowers (or florets) in the head come in two different kinds. First, "ray" florets which appear to have one petal, although it is actually five petals fused together. Second, "disc" florets, which are symmetrical and tubular with five equal-sized petals fused together. If one of these flower heads is broken apart, it is quite easy to see how the florets fit on to the receptacle at the top of the flower head stalk. In some species, the florets of either kind have both stamens and an ovary. Sometimes they only have male or female parts. Some species have florets of both kinds. Asters themselves, for instance, have ray florets round the edge and disc florets in the middle. Some, like the dandelion, have only ray florets and some, like Tansy, only disc florets. It is easy to see that the possible combinations make this a very complex family to understand. The way the flower heads are arranged may be an important feature to notice. In this book we refer to it as the inflorescence; some more technical books will call it the capitulescence because of the fact that these are not single flowers. The details of the flower head should be noted carefully and, also the types of leaf and how they are arranged. With this family, a magnifying lens is often a helpful tool.

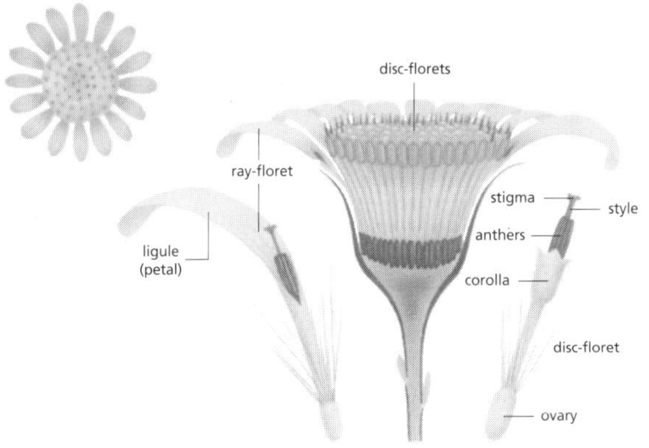

8. Trail Plant (Pathfinder) G5? S1

Adenocaulon bicolor **Hook.** **Aster Family**

[*Adeno* — gland; *caulon* — stem (Gr.)]

Trail plant is possibly so-called because of its distinctive bicoloured leaves (white-woolly below and green above) which could be used for marking a trail through the forest where it grows (Kozloff,1976). The generic name refers to the glands on the upper part of the stem. This is a very small group with only four species, the other three being found in Guatemala, Chile and Eastern Asia. This species is a western plant with its main distribution in the foothills and mountains from southern B.C. and Alberta to Idaho, Montana and California. It is also found in northern Michigan and northern Minnesota. Its presence in our area is hard to explain. The immature plants, with their rosettes of stalked, rather heart-shaped leaves, are somewhat similar to young Common Burdock. However, Burdock leaves have toothed edges, the flowers are quite different and the fruits have hooks.

Local Distribution and Habitat: Extremely rare. It is known nowhere else in Canada east of the Rocky Mountains. In 1895 it was collected on a "deeply wooded, rocky hillside" somewhere on the Cape Croker Indian Reserve. In 1994 it was rediscovered, not too far to the southeast, on the Meaford Tank Range (Canadian Armed Forces Base). It was found there again in 1996. The location is in deciduous woodland at the top of a steep slope near Georgian Bay. Habitat reasonably similar to that at the new site exists on the Reserve in an area that apparently has been little visited, so it is conceivable that it still occurs there also. In both cases the sites are on land which projects into Georgian Bay, giving a similar microclimate. The habitat cited for this species in standard texts is "moist, shady woods." Both these sites could fit that description.

Herbarium Specimen

DESCRIPTION

Height: 30-90 cm. Stem white woolly near base. Stalked glands in upper part (use lens).

Leaves: Alternate, mainly forming a rosette near the base. Green above, felty-white below. Long winged stalks. Triangular with large basal lobes. Untoothed edges with triangular lobes.

Flower Head Arrangement: Sparse. On a few slim branches from main stalk. ± 3 flower heads per branch.

Flower Heads: ± 5 mm wide, becoming wider as fruits form.

Bracts: Joined in a cup, oval lobes ± 2 mm long, reflexed in fruit. Deciduous at maturity.

Ray Florets: None.

Disc Florets: ± 3 mm long. **Whitish.** Only outer ring fertile.

Fruit: Club-shaped, glandular, 5-8 mm. No pappus.

Flowers: June to August (recorded from other areas).

9. Indian Plantain

G4G5 SC S3

Cacalia plantaginea **(Raf.) Shinn.**

Aster Family

Indian Plantain belongs to a North and Central American genus, with 50 species in the Americas and only one in Eurasia. They are much more prevalent in the west. Ontario is as far east as Indian Plantain can be found, and even there mainly in the Lake Huron-Georgian Bay counties. This plant was of interest to North American Indians as, like many wetland plants, it produces a tuberous root which is edible. Many plants with longitudinal leaf veins reminded Europeans of the common plantain and were named accordingly. Indian Plantain is normally found in wet areas, even in the prairies. It is a very striking plant, rising above most of the fen plants surrounding it, with rather silvery heads of flowers, and later with feathery seed parachutes. It is very highly dependent on the preservation of the natural integrity of wetlands with a high calcium content.

Local Distribution: Uncommon overall in Bruce. Very rare in Grey. Mostly fairly common along and near the Lake Huron shore, from MacGregor Point Park north to Dorcas Bay. Near the St. Edmunds-Lindsay Townline populations are found at least 6 kilometres inland. In the Oliphant area it is very common. It is also found at a few locations on the Georgian Bay shoreline north to Cabot Head, and inland including McCullough Lake and along the Saugeen River in Bruce. The plants at McCullough Lake in Grey County are a larger, more vigorous variation.

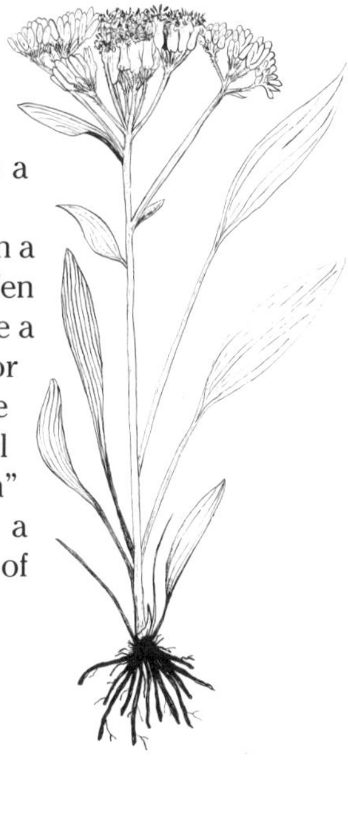

Local Habitat: Saturated to wet marly fens, where it is often a dominant or sub-dominant species. It can be abundant in tiny fen pockets in conifer-dominated swamps. Most of these fens have a sand substrate, but it also occurs in shallow fens on bedrock or wet pockets in alvars. Occasionally also found on inland lake shores, or on river shores or flats. Sometimes in very small quantity in cedar swamp habitat near fens. Also at two "dryish" open upland locations near the Saugeen River. Definitely a calciphile. This species is particularly vulnerable to the filling of wetlands and cottage development.

DESCRIPTION

Height: ± 25-120 cm. Stout, angled, striated stems.

Leaves: Lower leaves, long stalked. Blade ± 12 x 4 cm, elliptic. Few leaves up stalk, reduced size, becoming stalkless.

Flower Head Arrangement: Compound corymb ± 7-10 cm wide.

Flower Head: ± 1 cm wide, all discoid. **Whitish** 5-lobed corolla.

Fruit: Achene with parachute of white, rough bristles.

Flowering: Late June to early August, very rarely later. Note: In any year the majority of plants do not flower.

10. Hill's Thistle G3 S3

***Cirsium hillii* (Canby) Fernald** **Aster Family**

[*kirsion* — thistle (Gr.) First described by naturalist E.J. Hill (1833-1917)]

For a variety of reasons, thistles are not too popular with the human race. They probably evolved their prickles as a defence against grazing animals, which immediately brought them into conflict with early herders. There are several genera with over 400 species worldwide. Some of them are very common and widespread, but others, like Hill's Thistle, have evolved into very specialized habitats with a restricted geographical range. This is a species of, generally, open or semi-open areas, on sand or calcareous rock pavement. In Ontario, Hill's Thistle is only found in Bruce and Simcoe Counties and Manitoulin District. In the States it is found from Pennsylvania to Iowa and from Michigan to Minnesota and South Dakota. It is endangered in Ohio, threatened in Wisconsin and rare in Minnesota. The large purple flowers and the restricted habitat should make this thistle easy to distinguish from other commoner ones, although it can occur near Bull Thistle. However, this is considerably more prickly than Hill's Thistle when mature.

Local Distribution: Restricted to the Bruce Peninsula, where it could be considered uncommon. It is found at many locations within a narrow band along the west shore of the peninsula as far south as Sauble Beach. It can occur up to 6 or 7 km inland near the St. Edmunds-Lindsay townline. From Dorcas Bay to Little Pine Tree Harbour it could be considered common. It is also found on the east side of the peninsula on the Cape Croker Indian Reserve and Bears Rump Island.

Local Habitat: Relatively dry sites, both bedrock and sand. Usually in open to very open conifer or conifer dominated woodland, possibly sometimes in coniferous woods over old dolostone shingle or cobblestone beaches. White Pine, Jack Pine and Cedar are probably the most commonly associated tree species. Occasionally on open alvars including ones just above the Lake Huron high water line.

DESCRIPTION

Height: ± 20-70 cm. Hairy stems erect from long thickened, perennial roots.
Leaves: Lower leaves forming a basal rosette. Length 10-15 cm, width 2-4 cm. Tapering to a short stalk. Evenly lobed. Lobes spine-tipped. Green on both sides. Upper leaves smaller and stalkless.
Inflorescence: Usually a single flower head.
Flower Heads: 5-8 cm wide. **Purplish-pink.** Slightly sweet scented. Disc flowers only.
Bracts: Sticky, with a terminal prickle.
Fruit: Achene 4-5 mm. Bristly parachute.
Flowering: July, early August.

11. Pitcher's or Dune Thistle **G3 END S2**

***Cirsium pitcheri* (Torrey) Torrey and A. Gray** **Aster Family**

Pitcher's Thistle has a very restricted range. It is found only on sand dunes or sandy shores around Lake Michigan, Lake Huron and eastern Lake Superior, in Ontario, Michigan, Wisconsin and Indiana. It is already extirpated in Illinois. In Ontario it occurs in Lambton and Simcoe Counties and Manitoulin District as well as Bruce County. Manitoulin Island has most of the Canadian population. The rosettes of large deeply incised leaves reposing on the surface of dry sand are most striking. They are anchored deep in the sand by a strong taproot, and at flowering time will send up a tall shoot crowned with one or more whitish flower heads. One of the easiest places to see this plant in Ontario, at the present time, is Inverhuron Provincial Park. Its closure to camping for more than two decades may have helped Pitcher's Thistle to become entrenched on the shoreline dunes at present. It is to be hoped that now the park is re-opening again the management of the park will see that access is controlled (while still giving enthusiasts a chance to see it) so that this rare plant does not become extirpated in its last stronghold in this area. Apart from any aesthetic value, plants such as this play a very important role in stabilizing dune areas.

Local Distribution: Extremely rare. Known from Inverhuron Park since 1939 (at least), where it is still present. It was reported from Sauble Beach as far back as 1934 but has been gone since the 1960s. There was an unconfirmed report from Lyal Island in 1982. The only location where it is well established at present is Inverhuron Park. More than 80 plants were found in 2000, up from the count in 1981. Seriously threatened by dune buggies and ATV's as well as swimmers and sunbathers.

Local Habitat: Open dry sand dunes — both shoreline dunes adjacent to sand beaches, and slightly inland dunes (up to .5 km), where it may be adjacent to coniferous trees, including cedar, which provide some shade. Normally grasses or other herbaceous vegetation occur within 1 metre of these plants, nevertheless, they are found mainly on relatively barren sand on somewhat active dunes.

DESCRIPTION

Height: ± 20-100 cm. Felty-white flowering stalk arising from basal rosette.
Leaves: Very large. Deeply pinnately lobed. Lobes ± 50 x 7 mm, weakly spine tipped. Felty white, especially upper surface.
Inflorescence: Branched with a solitary flower head (or a few).
Flower Heads: Width ± 3 cm. Young flowers pale lavender. **White or cream** when mature. Bracts numerous, outer spine-tipped.
Fruits: Achene ± 6mm. Parachute of plumose bristles.
Flowering: July to August.

12. Lakeside Daisy, Stemless Rubberweed

G2 S2

***Hymenoxys herbacea* (Greene) Cusick** **Aster Family**

[*H.acaulis* (Pursh) Parker, *Actinea herbacea* (Greene) Robins.]
[*hymen* — membrane, *oxys* — pointed (Gr.); *herbacea* — not woody (Lat.)]

Lakeside Daisy is one of a group of twenty mainly western North American species of Rubberweed. This particular species is a Great Lakes endemic which has been found only in Bruce County and Manitoulin District in Ontario and in three of the Great Lakes states. Lakeside Daisy is endangered in Ohio, extirpated in Illinois, and known from only one location in Michigan. This group of plants are rather unusual among the Asteraceae. Instead of a feathery parachute on the achene there are five pointed, papery scales — hence the name *hymenoxys*. This is a perennial plant with a stout, persistent stem penetrating the ground from which grow a mass of fibrous roots. In spring it throws up a single bright yellow daisy-like flower head. The only plant likely to cause confusion is *Coreopsis lanceolata* which is found in similar habitat. Lakeside Daisy flowers are borne on hairy, leafless stems, and the petals of the rather oblong ray florets terminate in three equal teeth. Lance Leaved Coreopsis has leaves on the flowering stem, which is usually hairless or nearly so. The teeth on the Coreopsis petals are irregular and more than three.

Local Distribution: Known only from the most northern part of the Bruce Peninsula mainland, starting just south of the St. Edmunds-Lindsay Townline. Within this restricted area it is uncommon, although it can be locally very common. Several populations are found along or near the north shore from Cabot Head to north of Cyprus Lake. A few extensive populations occur well inland. The largest of all the populations, and one of the two found southwest of Highway 6, is now part of the Federation of Ontario Naturalists Nature Reserve where it is much visited and admired while being fully protected. This is the most southerly population in Ontario and was the first to be found on the Bruce Peninsula, in the early 1960s, even though the northern part of the Peninsula had been thoroughly botanized in the 1930s. This suggests that it has spread by seed dispersal from Manitoulin Island, where it is considerably more common, during the 20th century.

Local Habitat: Open, relatively dry dolostone. Either in alvars, or on bedrock (less often on boulders) on, or immediately adjacent to, the Georgian Bay shore. The alvar populations are generally larger, and are surrounded by woods, dominated by Jack Pine.

DESCRIPTION

Height: Flowering stem ± 15-20 cm. ± hairy.
Leaves: All basal, narrow, widest at top. Length ± 6 cm, centre width ± 6 mm. Tiny dark spots (use lens).
Flower Heads: Solitary, width ± 4 cm. ± 15 rays, 5-15 mm long, straight-sided, three rounded teeth at free end, **yellow**. Disc 8-20 mm wide. Bracts, oblong, rounded, hairy.
Fruit: Achene with five pointed, papery scales.
Flowering: May to June, rarely July.

13. Cylindrical Blazing Star

G5 S3

***Liatris cylindracea* Michaux** **Aster Family**

[*Liatris* — fringed (Lat.)]

Cylindrical Blazing Star is another plant with western affinities. Only thirteen of the thirty species of *Liatris* are found in eastern North America. The range of this species starts in Minnesota and reaches as far east as western New York State and southern Ontario. It is a plant of dry, open spaces or prairies, and may also be a remnant from a once wider distribution when the climate was warmer about 5000 years ago. The tall rods of feathery purple flower heads, characteristic of this group, make this plant impossible to miss when it is in flower. Although the flower heads are composed only of disc florets, the stigma has a long thread-like appendage. These, in combination, create the feathery appearance. Other species of Blazing Star are often grown as garden plants, but the flower spikes are elongate with numerous (± 100) small, stalkless heads. The flower heads in Cylindrical Blazing Star are stalked — at least the upper ones — and they are few in number.

Local Distribution: Extremely rare. Known only from two general locations a few kilometres apart, on the shore of Lake Huron northeast of Douglas Point. It was first recorded at Baie du Doré in 1966, but it is now believed to be extirpated at this location. Plants in the other location were first found in 1984. The stands in this second general location are all within one kilometre and consist of several hundred plants very close to Lake Huron (mostly within sight of it).

Local Habitat: Mainly in dry sand — largely in the form of open low dunes, but also in coniferous woods with a high Red Pine component.

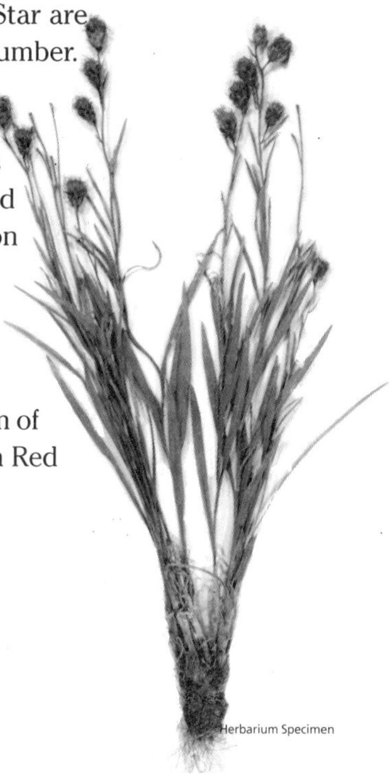

Herbarium Specimen

DESCRIPTION

Height: ± 30 cm. Stout hairless stems, sometimes branched.
Leaves: Linear ± 120 x 7 mm, reduced above. Firm texture.
Inflorescence: Raceme. Heads few or single. Short-stalked.
Flower Heads: 10-20 disc florets per head. Stigma with long appendage. **Pinkish-purple**. Bracts in 3 rows forming a cylindrical vase-like, receptacle with the stigma appendages spilling over.
Fruit: Achene with parachute of feathery, barbed bristles.
Flowering: July to October.

14. Roundleaf Ragwort

G5 S3

***Senecio obovatus* Muhlenberg** **Aster Family**

[*senex* — old man (Lat.), referring to the parachutes of white hairs on the fruiting head]

The Ragworts are a large and enormously varied group with over 1000 species worldwide, ranging from small weeds like Common Groundsel to sizeable shrubs and trees. In this area, we have five species, three native. Roundleaf Ragwort is very rare, in Ontario today found only on the Bruce Peninsula and Manitoulin Island. It is basically a southern species, ranging from Vermont to Florida and as far west as Arkansas and Texas. Thus it is well north of its main range here, with no continuity of distribution. This suggests that it might be a survivor from a once more widespread population. In some places it is called Running Groundsel as it reproduces from creeping stems. Like all Ragworts, the flowers are yellow. The petals are quite obvious but the flower head disc is comparatively small. As its Latin name suggests the leaves are slightly wider at the free end.

Local Distribution: Restricted to the Bruce Peninsula, where it is common overall becoming progressively less common to the south, rare south of Wiarton. The most southeastern point of its Canadian range is approximately 6 km ENE of the village of Shallow Lake. The NW point of its Canadian range is on Manitoulin Island and the bulk of the Canadian population is in St Edmunds and Lindsay Townships.

Local Habitat: Upland woodlands (unlike other Ragworts). Sometimes in cedar swamps, or in openings (such as road verges) adjacent to woodlands. Most often on thin soil over bedrock, sometimes on loam, or talus, or other forms of dolostone. Absent in deciduous swamps and normally absent on sand. Woodland habitats are usually coniferous to mixed forest, but it also occurs in deciduous forest. Associated with a number of tree species, especially White Cedar (first), White Birch, Balsam Fir and Jack Pine. It commonly forms patches, sometimes with hundreds of leaf rosettes; very often none will bloom in a given year. It appears to hybridize with Balsam Ragwort, very rarely.

Herbarium Specimen

DESCRIPTION

Height: ± 40 cm. Weak, erect, hairless stems. Runners above or below ground.

Leaves: Basal rosette, spoon-shaped, winged stalk, ± 7 x 2 cm (with stalk), often purplish, edge with small rounded teeth. Few stem leaves — smaller, stalkless, oblong, toothed.

Inflorescence: Flat-topped corymb with ± 8 long-stalked flower heads.

Flower Head: Width ± 2 cm. 8-12 **Golden-yellow** rays, 5-8 mm long. Disc ± 5 mm wide. Bracts numerous, narrow, sharp-pointed, ± 4 mm long.

Fruit: Achene with white parachute.

Flowering: (Late May) June, early July.

15. Houghton's Goldenrod **G3 S2**

Solidago houghtonii **Torrey & A. Gray** **Aster Family**

Most people can easily recognize a Goldenrod, especially the familiar Canada Goldenrod group which is extremely invasive in wasteland and old fields. However, in our area we have, at present, fifteen different kinds of Goldenrods (Semple et al, 1999). Of the three rare Ontario species that are listed for this area only two have been found recently. Houghton's Goldenrod is another of the Lake Huron endemics present in this area, which are found only around the Lake Huron shore in Michigan and Ontario. Its flower head arrangement is flat topped and there are only two others of that form. Grass Leaved Goldenrod is very common but does have very narrow grass-like leaves. Ohio Goldenrod is also flat topped but has a characteristic basal rosette of quite wide, long-oval leaves. It grows in wet places even when it is on open rock. Houghton's Goldenrod has narrower leaves but not narrow enough to be called grass-like. It grows in drier areas. The extreme rarity of this species in our area makes it very vulnerable to extirpation from even minor changes to the environment in which it grows.

Local Distribution: Exceedingly rare. Only known from the Cabot head region on Georgian Bay, between the cliffs and the shore.

Local Habitat: Dolostone or limestone pavements. (On sand dunes in other areas). It was once seen growing with Ohio Goldenrod at Cabot Head.

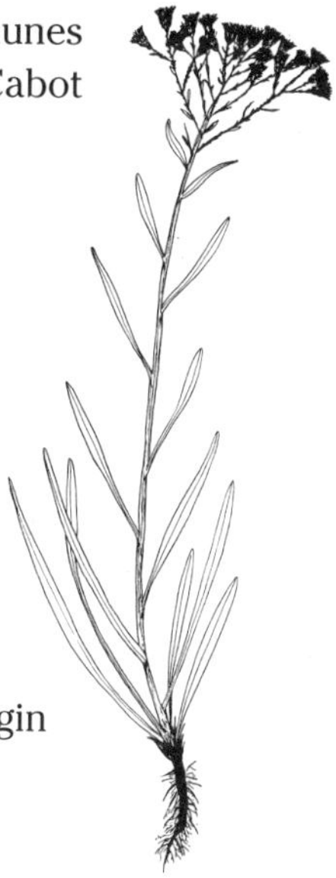

DESCRIPTION
Height: ± 30-50 cm.
Leaves:
Basal: Ovate. Tapering to a long winged stalk. Margin entire. Hairless. Forming rosettes.
Lower Stem: Linear-oblanceolate. Tapering to a winged base. Margin entire. Hairless.
Upper Stem: Similar but reduced and stalkless.
Flower Head:
Diameter: ± 10 mm.
Rays: 6-12 Pale to bright yellow.
Disc Florets: 8-15, **yellow**.
Bracts: Linear, fringed in upper half. Faintly striped.
Flower Head Arrangement: Flat-topped corymb. Branches usually sparsely hairy with a few leafy bracts. Few Heads, 2-18.
Flowering: August to September.

16. Ontario Goldenrod G3? S3?

***Solidago simplex* Kunth ssp. *randii* (Porter) Ringius var. *ontarioensis* Ringius**
[*S. glutinosa* Nutt. *S. spathulata* DC. *S. randii* Porter, *S.racemosa* Greene] **Aster Family**

Solidago simplex is anything but simple. It has undergone many confusing name changes, but with modern techniques, botanists now split it into three varieties, only one of which is found in this area (Semple et al, 1999). The range of Ontario Goldenrod does extend north of here into Manitoulin Island and along the shoreline of Algoma District. It is an endemic variety of the Lake Huron and Lake Superior shores and is found nowhere else. Its habitat requirements and the fact that it flowers much earlier than any other Goldenrod make it fairly easy to identify. It is also generally smaller than the later Goldenrods we are used to seeing in the Fall. This is very much a shoreline plant and this coupled with its rarity, would make it very vulnerable to shoreline development.

Local Distribution: Uncommon. Found near the lake, mainly well to the north, particularly on the Tobermory Islands. Most frequent on the Georgian Bay side of the Bruce Peninsula, where it occurs at least as far south as Lion's Head. It also occurs on the west side mainly near the tip, but it is known south to Red Bay. It is has also been reported as far south as Bruce Township.

Local Habitat: Dolostone; shoreline rocks or clifftops near the lake. Rarely on sand.

DESCRIPTION
Height: 5-50 cm.
Stems: Hairy towards the tip, slightly sticky. Downy on flower head branches.
Leaves:
Basal Rosette & Lower Stem: Narrow, oblanceolate. Narrowed to a long stalk. Margins toothed.
Upper Stem: Similar, size reducing up stem. Sometimes sticky.
Flower head:
Diameter: ± 6 mm.
Rays: 7-16. **Yellow** conspicuous.
Disk Florets: 6-30.
Bracts: Ovate, blunt, often sticky. Distinct midrib.
Flower Head Arrangement: Panicle. Branches simple racemes, arising in the axil of a leaf. Few flowers. Plants in poor conditions have short branches with very few flowers.
Flowering: Late June to August. Sometimes a little earlier or later.

17. Hairy Puccoon, Plains Puccoon G4G5 S3

***Lithospermum caroliniense* (Walter) McMillan var. *croceum* (Fern.) Cronquist**

[*Lithospermum croceum* Fern.] **Forget-me-not Family**

[*lithos* —stone, *sperma* — seed (Gr.)]

One of the characteristics of this family is that they are most often hairy, and Hairy Puccoon, as its name suggests, is no exception. The *Lithospermums* are so called because their seeds do look like little shiny white stones. The bright orange-yellow petals of this species, against the green leaves and sandy background of its habitat, make this flower hard to miss. If you go to MacGregor Point Provincial Park the first weekend in June you will find it growing on a dry, open sandy flat not far from the shore. The key to finding this plant seems to be sandy soil. There are two varieties of *Lithospermum caroliniense*; the one we have here, var. *croceum*, extends through southwestern Ontario into the Midwest as far as Oklahoma. The other variety var. *caroliniense* has more of an Atlantic coastal distribution as far south as Florida and west to Texas. This is not, in general, an endangered species but neither is it very common. Sandy habitats are often very fragile and, especially along coasts, subject to development.

Local Distribution: Rare. Known only from the Lake Huron shore in Bruce County from Chiefs Point Indian Reserve to the village of Inverhuron; most prevalent within 2 km of the mouth of the Sauble River. Even in this narrowly defined strip it would still be considered very uncommon. It should be looked for in Huron Township on sand at the shore of Lake Huron, but the density of cottage development and recreational use along this coast is probably detrimental to its existence.

Local Habitat: Dry sand dunes, including very low ones. Most often within sight of Lake Huron, never more than 1 km inland. Mostly in open areas but some populations are in very open woods, mostly predominately coniferous, where sunlight reaches the ground.

Herbarium Specimen

DESCRIPTION

Height: ± 30 cm. Stems erect, clustered from a stout root.

Leaves: ± 3 cm long, linear, rough hairy.

Inflorescence: Leafy-bracted cyme.

Flowers: 5 **Orange-yellow** petals joined in a tube. 5 sepals, 5 stamens.

Fruit: Shiny ivory-white nutlets ± 3 mm long.

Flowering: Late May, June.

18. Lake Cress

G4? S3?

***Neobeckia aquatica* (Eaton) Britton**

Mustard Family

Armoracia lacustris (A.Gray) Al-Shehbaz & V.Bates

[*aquatica* — of water; *lacustris* — inhabiting lakes (Lat.)]

This is a true aquatic plant, growing rooted in shallow water and, like a number of aquatic plants, has two kinds of leaves; feathery ones under the water and more substantial ones rising above it. This is a very small group with only four species in the northern hemisphere. They all have the typical, four-petalled cress flower like their dry land relatives. Most aquatic plants, including Lake Cress, are very susceptible to environmental changes, especially pollution and wetland drainage. It appears to be declining over much of its range in eastern North America. It has been reported at some time from eighteen counties in Ontario, but apparently it has disappeared from a number of them. Nevertheless, there are known to be extant populations in Renfrew County and Manitoulin District, as well as Bruce and Grey Counties.

Local Distribution: Rare. Various locations along the Sauble River. In or near the Beaver River below Kimberley. Known from Shallow Lake and other locations on the Bruce Peninsula at least as far north as Johnston's Harbour. It may have been found in other township units on the Bruce Peninsula that are not recorded on the map.

Local Habitat: Rivers and streams, usually sluggish water. Also known from "marshes" (Krotkov, 1940) and a shallow lake.

DESCRIPTION

Height: 30-60 cm. Branching.

Leaves: Submerged leaves: Pinnately dissected into filiform segments. Above Water: Lanceolate to narrow oblong, ± 5 cm long. Dentate. Often few or none

Inflorescence: Raceme. Emerging from water.

Flowers: Petals 4, obovate, **white**. Stalked.

Fruit: Ellipsoid, ± 6 mm long, persistent style ± 3 mm.

Flowering: Summer.

19. Stiff Gentian — Ague Weed, Gall Weed

G5 S2

Gentian Family

***Gentianella quinquefolia* (L.) Small** [*Gentiana quinquefolia* L.]
[*Gentius* — king of Illyria, discoverer of medicinal properties of Yellow Gentian]

The medicinal properties of the roots of the southern European *Gentiana lutea* — Yellow Gentian have been known for a very long time. They were very successful tonic medicines, probably due in part to their intensely bitter taste! Judging by the older English names of this North American plant, this tradition must have been continued by early settlers. How effective this was, we do not know. The old genus *Gentiana*, including almost all of Canada's gentians, has been split into three. *Gentianella* is found around the northern hemisphere, mainly in areas with cold climates, to the north or in the mountains. Plants in this genus are relatively delicate compared with *Gentianopsis* — the Fringed Gentians. The third genus which retains the name *Gentiana* has the plants with relatively closed flowers such as the Bottle Gentians.

Stiff Gentian is eastern North American, generally found in warmer climates than any other Canadian species of its genus. In this area it is at the northern edge of its range, which extends from New England to Georgia and as far west as Minnesota and Arkansas.

Local Distribution and Habitat: Extremely rare. Known only from two places on the Saugeen Indian Reserve, both on moist roadsides with a hint of shade. It was present at one sandy site from 1973 (at least) until the end of the 1990's. At peak population there were up to 100 plants. It was first seen at the second site a little later but there were never more than a fraction of that number and there was none by the end of the 1980's.

Throughout its range, the habitat seems to vary from dry to moist soil, and from wooded to open areas. In this area, it should probably be looked for along roadsides.

DESCRIPTION
Height: ± 10-30 cm. Erect stems.
Leaves: Sessile, lance-ovate, sharp pointed. 3-7 parallel veins, ± 2.5 cm long.
Inflorescence: Terminal cluster, occasionally solitary flowers in leaf axils.
Flowers: Very small. **Pale pinkish-purple**. Short-stalked, calyx lobes linear. Petals joined in a tube, 4x length of calyx.
Fruit: Stalked capsule.
Flowering: August-October.

20. Stiff Yellow Flax G? S2

Linum medium* (Plan.) Britton var. *medium **Flax Family**

[*Linum* — name for flax in classical Latin]

The typical flax plant that is used for fibre and for oil from the seeds may often be seen in fields, especially in central Canada, making a sea of ethereal blue when in full bloom. It also occurs as a wild plant in many places, although its origin was in Europe. There is one native blue flax in western North America, but all the other native species have yellow flowers. However, they have many of the same characteristics of the cultivated flax with stiff fibrous stems, and small, narrow leaves, as well as flowers with five petals, sepals, stamens and five sections in the capsule which produce numerous small seeds. There are two native yellow species in this area (see No. 21) which are both considerably smaller than cultivated flax. They are both provincially rare and near the northern edge of their range. The variety *texanum* of Stiff Yellow Flax is found in southwestern Ontario and from New England south to Florida, west in the mid-western states, and as far southwest as eastern Texas, but the variety *medium* which is found here is a Great Lakes endemic, only occurring near western Lake Ontario, Lake Erie and Lake Huron.

Local Distribution: Rare overall. Along the Lake Huron shore in Bruce County, including several islands, from the north boundary of Eastnor Township south to Macgregor Point Park and probably as far as Baie du Doré. Mostly, it is within sight of Lake Huron, but sometimes up to 1 km or more inland where there is suitable habitat. Within that stretch it could be said to be uncommon rather than rare. There has been an unconfirmed report from the Cape Croker Indian Reserve on the east side of the Bruce Peninsula.

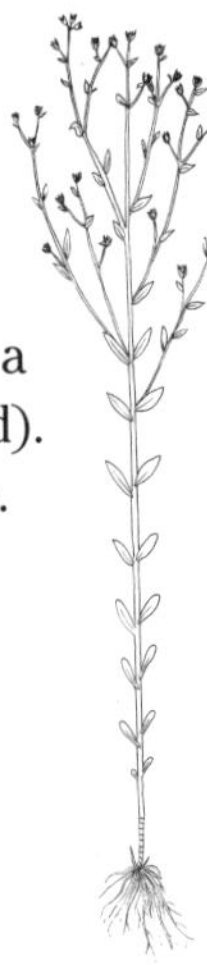

Local Habitat: Shorelines — dolostone, or sandy, or sometimes muddy, marly sandy meadows and less wet portions of fens. Once in a marly roadside ditch and in Cape Croker at a prairie alvar (not confirmed). Often a colony appears in a location and then disappears after a few years.

DESCRIPTION

Height: ± 15-30 cm. Stiff, erect, grooved stem.

Leaves: Narrowly lanceolate 10 x 3 mm. Smaller above. Linear. Greyish-green. Glandular dotted.

Inflorescence: Small corymb with few branches, 2-5 flowers per branch.

Flower: 5 petals 5-8 mm long, **Yellow**. 5 sepals sharp-pointed, gland toothed.

Fruit: Capsule, flattened-globose, 2 mm long. Borne in a cup formed by the sepals.

Flowering: July to August. This is a perennial, spreading by thin rhizomes.

21. Grooved Yellow Flax **G5 S3**

***Linum sulcatum* Riddell** **Flax Family**

[*sulcatum* grooved (Lat.)]

The name of this plant refers to the stem which is angled and somewhat winged, thus it appears to be grooved. A pair of minute dark glands at the base of the leaf is also characteristic, although a lens will be necessary to see them clearly. The outer sepals in this species are glandular-ciliate. The flowers, too, tend to be slightly larger than in Stiff Yellow Flax. This is somewhat of a more western species, although it is found in New England. Its range extends well into Texas, but it is not found in the southeastern states.

Local Distribution and Habitat: Extremely rare. It is established on the Cape Croker prairie alvar, on limestone or calcareous shale, where it was still flourishing in the summer of 2000. Also on nearby Barrier Island (alternatively known as Rabbit Island) on open bedrock pavement.

Herbarium Specimen

DESCRIPTION

Height: ± 20-45 cm. Stiff stems.

Leaves: Linear, sharp-pointed, 2 cm x 3mm. Pair of minute dark glands at base.

Inflorescence: Panicle.

Flowers: 5 petals, 5-10 mm long, **yellow**. 5 sepals 3-6 mm long, conspicuously glandular-ciliate.

Fruit: Ovoid-globose 2-2.5 mm. Borne in a cup formed by the sepals. This is an annual, depending for survival on dispersal of seeds every year.

Flowering: (July) August (September).

22. Pine Drops

G5 S2

***Pterospora andromedea* Nutt.**

Indian Pipe Family

[*ptero* — winged, *spora* — seed (L.)]

This small family, with ten of the twelve species in North America, is very highly specialized. Many plants, especially trees and orchids, have an association with soil fungi which helps provide them with the nutrients they need. Indiscriminate use of herbicidal chemicals can be counter-productive by killing these essential organisms. The Monotropaceae have become so dependent on mycorrizhal fungi to extract nutrients from the organic matter in the soil, that they have ceased to produce chlorophyll, and so are no longer green. Most people have seen the eerie white of Indian Pipe, silhouetted against the forest floor. Pine Drops is related but it has a purplish-brownish colour. This, of course makes it less easy to spot in the conifer dominated forests where it grows. The leaves in these plants are reduced to scales, because they do not need them to catch light for photosynthesis. The single stems have scale leaves at the base, and bear a spire of small bell-shaped flowers, which betray the relationship of this deviant group to the Pyrolas or Wintergreens. This is an extremely rare plant in Ontario, although there are a number of historical reports.

Local Distribution: Very rare. Known only from the Bruce Peninsula where it has been seen on four different occasions between the mid-1960's and 2000. It has been found at three or four locations, one at Red Bay and the others towards the north shore of the peninsula. At least two of the sightings were of a single plant. It has not been seen again at these locations, so it may not establish itself easily.

Local Habitat: All the occurrences have been in woodland. The plant at Red Bay was in a lowland wood adjacent to a fen and dominated by cedar. At least two of the sightings to the north have been in upland mixed or coniferous forests.

DESCRIPTION

Height: ± 40 cm. Single, glandular-hairy purplish-brown flowering stems, connected by rhizomes.
Leaves: Reduced to scales.
Inflorescence: Raceme ± 15 cm.
Flowers: Calyx 5-lobed. Corolla bell-shaped, pinched at 5-lobed mouth. 6-7 mm, usually **reddish.** Outer sepals not glandular-ciliate. Stamens 10.
Fruit: A capsule, five compartments, tiny winged seeds.
Flowering: Summer.

23. Broad Leaved Hawthorn — G3G5 S2S3

Crataegus corusca **Sarg.** — **Rose Family**

[*kratos* — strong, (Gr.), *corusca* — to sparkle or vibrate (Lat.)]

The Rose Family is a very large one with 3000 species world-wide, many of which are of economic importance. Most of our temperate fruits belong in this family. The Hawthorns (*Crataegus*) are a widespread, largely North American, group of small trees or shrubs whose main range extends into the Andes. They are characterized by their thorns and their fruits — "haws." The haw is a pome, usually red, and similar to a tiny apple, with the fleshy part derived from the stem and enclosing core and seeds. The fairly small flowers are simple with 5 petals, usually white, and numerous stamens. These are very much successional species invading disturbed sites such as abandoned fields, old orchards and regenerating forest. This genus is one of the most complicated and one of the most studied. At one time 1000 species were described but many of them turned out to be crosses of two species, or in some cases types that arose without benefit of sexual reproduction. The number of species today is between 100 and 300 depending on who is counting! This particular species was originally a cross between Scarlet Hawthorn (*C. coccinea*) and Downy Hawthorn (*C. mollis*) but it is sufficiently well established and distinct to be recognised as a species in its own right. It displays hybrid vigour by being bigger than its ancestors and with wider leaves. It has been reported from a number of Ontario counties, mostly in the southwest, the furthest east being Peterborough. Precise identification needs an expert, so it is probably often overlooked among the mass of other Hawthorn species.

Local Distribution: Uncommon or rare. Known from approximately four locations in Keppel Township and three further south in Grey County. Not known from Bruce County.

Local Habitat: Abandoned pastures being invaded by woody species, roadsides, young deciduous woods etc. In some cases, there are substantial stands of this species, not just isolated trees.

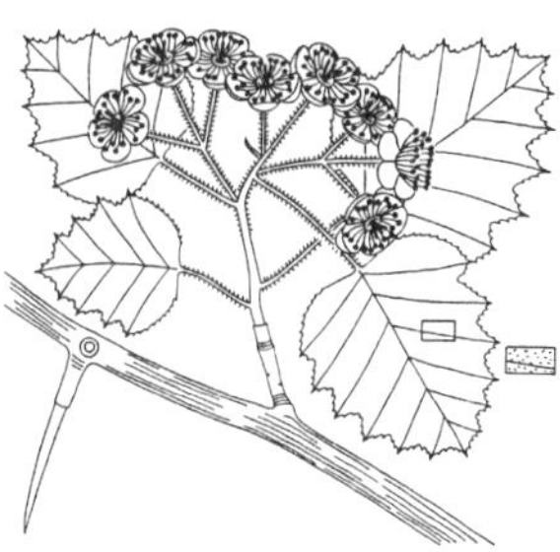

DESCRIPTION

Height: Up to 6 m. Tree with ascending branches. Spines 2-5 cm, sparse.

Leaves: Broadly ovate, large teeth, serrate edge. Smooth or hairy above, velvety below. Leaf ± 5 cm long, leaf stalk ± 3 cm long

Inflorescence: Corymb 8-12 flowers, ± 2 cm diameter. Stalks furry.

Flowers: ± 2 cm wide. 5 **white** petals. ± 20 pink stamens. 5 sepals joined at base.

Fruit: Bright red, ovoid — widest near top, Calyx persistent, erect, 4-5 nutlets.

Flowering: Late May, June.

24. Cauwanagha Thorn G3? S1

Crataegus suborbiculata **Sarg.** **Rose Family**

[*suborbiculata* — almost round (Lat.)]

The Latin specific name refers to the leaves which are sometimes almost round. *Crataegus suborbiculata* seems to have an affinity for limestone areas. In this area, it is a large thorny shrub, although it can develop into a 7 metre tree under ideal conditions of soil and climate. Its range extends from New England through the St. Lawrence valley to the lower Great Lakes. Most of the reports in Ontario are from the southwest — Lambton, Middlesex and Elgin counties. Cauwanagha Thorn is fairly similar in appearance to Dotted Hawthorn (*C. punctata*), which is the most common hawthorn in this area.

Local Distribution: Very rare. Known only from a few locations in Keppel Township. Two populations are definitely known to be in the Big Bay area. The specimens in this area are said to be a distinct variation on the regular Cauwanagha Thorn, but the populations are very isolated, being much further north than any others in Ontario. They have probably been separated from the southern populations for a long time. The genetic variability of Hawthorn species would make such changes very likely.

Local Habitat: Abandoned upland pastures, among other hawthorns, escaped pear trees and other woody species. One of these populations comprises many hundreds, if not thousands of individuals of this species.

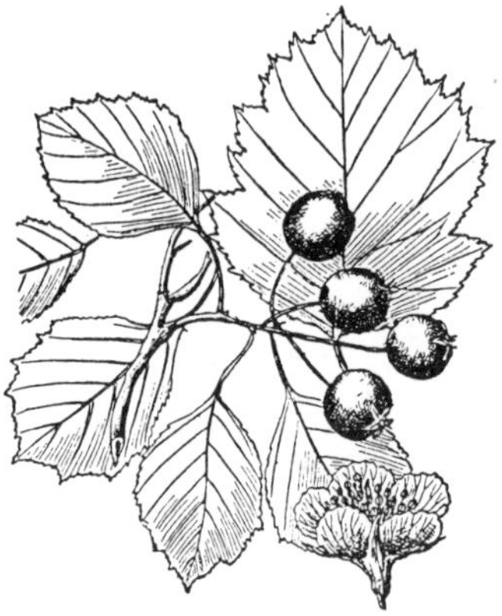

DESCRIPTION

Height: Up to 4 m. Large shrub. Spines ± 3 cm.

Leaves: Ovate-orbicular, ± 4 x 3 cm. Shallowly toothed, edge serrate. Dull green, not hairy.

Flowers: 2 cm wide. **White**. 5 petals, ± 20 stamens, usually pink. Sepals narrow.

Fruit: Globose with a persistent calyx. Dull red. ± 1.25 cm diameter. 3-5 nutlets.

Flowering: June.

25. Large White Flowered Ground Cherry

G3? S3?

***Leucophysalis grandiflora* (Hooker) Rydberg** **Nightshade Family**

[*Chamaesaracha grandiflora* (Hook.) Fern.]

[*leucophysalis* (Gr.) — white bladder (inflated calyx); *grandiflora* (Lat.) big flower]

This striking plant is most closely related to *Physalis*, one species of which is the familiar Chinese Lantern Plant. The orange "lantern" that encloses the berry is actually the calyx. *Leucophysalis grandiflora* is in a separate genus because, although its calyx persists, it does not close to form a lantern. It also turns whitish, not orange, and the berry almost completely fills it. In this family the petals are joined and often form a rather star-shaped corolla. The shape is reminiscent of a bindweed. Most of this group produce berries and the tomato is one of the best examples. This species is found from Quebec to Saskatchewan and in the States from Vermont to Minnesota, so it is in the middle of its range here. It is a rather short-lived perennial, found on disturbed sites, and does not seem to be too common anywhere, tending to appear sporadically.

Local Distribution: Very rare. Known from six sites in recent decades — mainly in the Sauble Beach area or towards the tip of the peninsula. It was considered "not uncommon" on the peninsula in the 1930's and may have been found in any of the townships between St. Edmunds and Amabel at that time. There is one recent record from Holland Township in Grey County. Not reported from Grey-Bruce since the 1980s.

Local Habitat: Dry open sites with some disturbance, often on roadsides, Sauble Beach sites on sand. To the north it was found in disturbed non-sandy sites such as gravelly road shoulders. Holland Twp. habitat not known. Sites recorded in the 1930s were "in dry sandy or gravelly soil."

Herbarium Specimen

DESCRIPTION

Height: Up to 1 m. Bushy from a taproot.

Leaves: Ovate, ± 8 cm long. Sharp pointed.

Inflorescence: Group of 2 to 4 in leaf axils.

Flowers: Stalked. 5-pointed cup-like calyx. Petals papery, joined, 5 very shallow lobes. Width 4-5 cm. **White, centre pale yellow.**

Fruit: A many-seeded berry.

Flowering: June to July.

26. Spiny Fruited Coontail **G4? S3**

***Ceratophyllum echinatum* A.Gray** **Coontail Family**

[*keras* — horn, *phyllon* — leaf (Gr.); *echinatum* — spiny (Lat.)]

Coontails are also called Hornworts because their branching is reminiscent of a spread of antlers. World-wide, there are only six species with two in North America. However, the Common Coontail is a cosmopolitan species, whereas the Spiny Fruited Coontail is only found in North America, extending into tropical America. These are rootless aquatics that live totally submerged. The stems are elongate and branch freely, forming large masses. The densely branched, feathery ends of the stems are like a Raccoon's tail. The most conspicuous difference between the two North American species lies in the leaves. In the Common Coontail (*C. demersum*) no leaves are forked more than twice, In the Spiny Fruited Coontail there are always some leaves forked more than twice. The fruits also differ. In the Common Coontail, there are only two basal spines, but in the Spiny Coontail, there are several marginal spines in addition to the basal ones. These are plants of quiet water in ponds and streams. Spiny Fruited Coontail has been reported from many parts of Ontario but it is probably frequently overlooked and only noticed when a comprehensive study of an area is done by a competent botanist. In Ontario it occurs mainly on the Precambrian Shield but it is also found off the shield on Manitoulin Island, in Essex County, and Waterloo Region as well as in Bruce County.

Local Distribution: Known only from Kinlough Creek and from the Sauble River at Allenford, where it runs through farmland. Both sites recorded in the late 1980's.

Local Habitat: Kinlough Creek is a sluggish stream in deciduous swamp habitat in Greenock Swamp.

Herbarium Specimen

DESCRIPTION

Stems: Elongate, brittle.

Leaves: In whorls. Palmately 2-3 (-4) times forked into filiform segments with tiny teeth. [Note: In *C.demersum* all leaves are twice forked. In *C. echinatum* some leaves will be forked more than twice.]

Flowers: Small and inconspicuous, no petals. Single in axils of leaves. Upper ones staminate with ± 12 stamens. Lower ones pistillate, single ovary with a filiform style.

Fruit: Achene with spines.

Flowering: Not known.

27. Blueleaf Willow

G4 S3

***Salix myricoides* (Muhl.) J.Carey** var. *myricoides*

Willow Family

[*S.glaucophylloides* Fern.]

[*salix* — willow in classical Latin, *myricoides* — resemblance to *Myrica* (Bayberry, Sweet Gale)]

The Willow family is a large one with over 300 woody species of shrubs and trees, mostly found in the temperate region of the northern hemisphere but some species extend into the Arctic beyond the tree line. This family has been known from ancient times and among many nations for the anti-inflammatory properties of its bark, which contains salicylic acid — from which the active ingredient of Aspirin is derived. There are many species of willow, twenty being listed for Grey and Bruce Counties. The clue to having located this species would be finding a small shrub growing on sand along Lake Huron which has leaves with a waxy powder below. The twigs are shiny, yellowish to dark brown. The erect catkins appear just before the leaves come out. Often it will be found with the related Sand Dune Willow (*S. cordata*) but this has much wider and hairier leaves and the young twigs are grey and hairy.

Blueleaf Willow is not a very distinctive or conspicuous species, so the best way to be sure you have found this is to check with an expert. The Ministry of Natural Resources or your local Field Naturalists Club will help.

Local Distribution: Rare. Restricted to the shore of Lake Huron in Bruce County from Lurgan Beach to the north part of the Oliphant area. The shoreline dunes at Sauble Beach are probably the easiest place to find it. It occurs there with Sand Dune Willow.

Local Habitat: Sand, mostly dunes. Largely in view of Lake Huron but sometimes slightly inland. Rarely in semi-wooded situations inland.

Herbarium Specimen

DESCRIPTION

Height: ± 2 m, generally shorter.

Leaves: Alternate, deciduous ± 7 x 1.5 cm. Lanceolate to elliptic, pointed, margins toothed. Upper surface dark green. Lower surface whitish to bluish waxy.

Flowers: Staminate catkins (on separate bush) ± 2 cm long, stamens 2. **Silver,** golden when mature. Scales brownish-black, hairy. Pistillate catkins ± 3 cm long. **Greenish.** Scales brown, papery, hairy.

Fruits: Slender capsules on female catkin, releasing seeds with tufts of silky hairs.

Flowering: May, as leaves begin to appear.

28. Heart Leaved Alexanders **G5 S1**

Zizia aptera **(A.Gray) Fern.** **Parsley Family**

[Alexanders probably a corruption of *olus* pot-herb; *atrum* black.]

Heart Leaved Alexanders is a typical member of the Parsley or Carrot Family, which is characterized by the umbrella-like arrangement of the small flowers. Plants in this group are found world-wide and many of them are prized for their edible roots and stems, or for their flavourful leaves or seeds which are used as herbs. The genus *Zizia*, although named after a nineteenth century German botanist, I.B.Ziz, is exclusively North American and has only four species. The common name "Alexanders" seems to have been inherited from a European plant which originated in the Mediterranean area. Also known as Horse Parsley, this was a precursor of celery for culinary use. The flowers are greenish yellow, so when early settlers encountered the very similar *Zizia aurea* in eastern North America they called it Golden Alexanders because of its brighter colour. This species *Zizia aptera* is distinguished by its heart shaped basal leaves. The basal leaves and the thrice divided stem leaves make this plant easy to distinguish from the other yellow-flowered member of this family which will be encountered in this area. Wild Parsnip is common everywhere, but it has pinnately divided leaves.

Heart Leaved Alexanders is found across the eastern United States, and throughout the prairies as far as Oregon and British Columbia. The main area of distribution is south and west of Ontario. It is extremely rare in this province. The only locations reported in the last twenty years are in northwestern Ontario and along the Lake Huron shore in Bruce and Huron Counties.

Local Distribution: Very rare. Known from only three sites in Bruce County, along the west shore of the peninsula adjacent to Lake Huron, between Southampton and Oliphant. Recorded from 1948. Still present. In 1995 hundreds of plants were recorded at a site near Chiefs Point, one third blooming.

Local Habitat: The two populations seen recently were in openings in, or on the edge of, conifer dominated woods with a high proportion of species such as cedar and tamarack, more or less on dolostone bedrock, near Lake Huron.

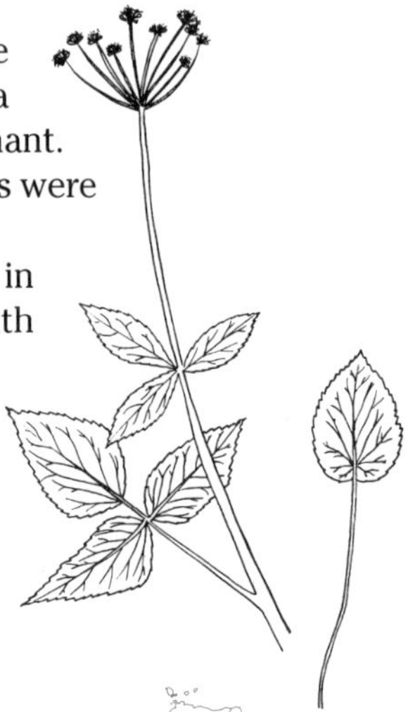

DESCRIPTION

Height: ± 30-80 cm

Stem: Hairless from a cluster of thickened roots.

Leaves:

Basal: Slightly heart shaped, toothed, long stalked.

Stem: Divided into 3 toothed leaflets.

Inflorescence: Compound umbel, main stalks of mature umbel ± 5 cm.

Flowers: Numerous, tiny, **bright yellow**, 5 petals, 5 sepals.

Fruit: Dry fruit splitting into 5 seeds, 1-2 mm long.

Flowering Time: Early summer.

29. Ginseng G3G4 END S3

***Panax quinquefolius* L.** **Sarsaparilla Family**

[*Panakos* (Gr.) — all healing; *quinquefolius* (Lat.) — five leaves (leaflets)]

Ginseng is two Chinese words meaning "man-image," in reference to the forked root. In China *Panax ginseng* was so highly prized for medicinal purposes that it was a monopoly of the Manchu emperor. When Europeans discovered a similar plant, *Panax quinquefolius,* in North America they jumped at the chance to make a fortune. Shiploads of it were sent to Canton with the result that, from being a common plant in eastern North America, it became a rarity. Fortunately, it was discovered that it could be cultivated, reducing the pressure on the wild plants, but, there are still those who collect it from the wild claiming that it is "better" than the cultivated kind. The Sarsaparilla family has several things in common with the Parsley family. The use as herbs or flavouring is one, and the arrangement of very small flowers in an umbel is another, but in this family the top of the umbel is rounded, often forming a ball rather than a flat topped umbrella. Ginseng might be confused with the two Sarsaparillas (*Aralia nudicaulis* and *Aralia hispida*), but these both have pinnately compound leaves, with short stalked or sessile leaflets. Their flower heads are on a separate stalk and tend to be larger and globular. Their fruits are black. Dwarf Ginseng (*Panax trifolius),* also found in Bruce County, is much smaller — max. 20 cm tall — with 3 (usually) sessile leaflets. The flowers are white or pink tinged and the fruits are yellow. Ginseng is an eastern North American plant on the northern edge of its range here. There is probably more of it in Grey County than most parts of Ontario, partly because there is more remnant forest. In many areas, logging and habitat destruction as well as harvesting pose a threat to its existence.

Local Distribution: Rare in Bruce. Uncommon in Grey. Exhibits some affinity for the Niagara Escarpment and its vicinity. Presumably, common in Grey County prior to European settlement.

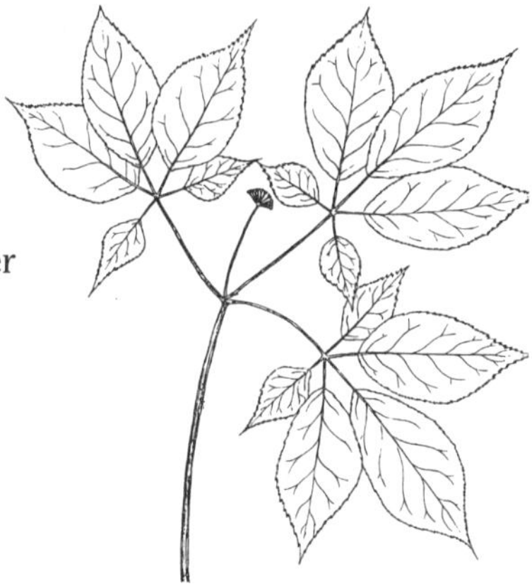

Local Habitat: Rich, upland deciduous forest, usually dominated by Sugar Maple, and usually under closed canopy. Usually in deep leaf litter, on loam (often near dolostone boulders but sometimes at non-rocky sites), dolostone bedrock (often with a thin layer of rich loam), or dolostone talus.

DESCRIPTION

Height: 10-60 cm.

Leaves: Usually 3 compound leaves, palmately divided into 5 leaflets. Leaflets oval, tapering, toothed.

Inflorescence: Simple umbel, small, round topped. On a stalk arising at base of leaf stalks.

Flowers: Small, **Greenish-white**, 5 petals, 5 sepals.

Fruit: Flattened berry with 2 or 3 seeds. **Red.**

Flowering: Early summer.

30. Marsh Valerian, Swamp Valerian G4G5 S2

***Valeriana sitchensis* Bong ssp. *uliginosa* (Torrey & A.Gray) F. Meyer**
[*Valeriana uliginosa* (Torrey & A.Gray) Rydberg] **Valerian Family**

Valerian is a name which was given to members of this family in Europe. It is an old personal name possibly dating back to Roman times. It was not unusual to personalize plants in this way in Mediaeval times. These names were then imported into the New World and applied to similar plants here. Characteristically, the Valerians have rather small flowers arranged in a complex corymb-like head. Marsh Valerian is easy to separate from the European Valerian *Valeriana officinalis* or Garden Heliotrope, which often escapes from gardens. In Garden Heliotrope all the leaves are pinnately divided, but in Marsh Valerian the basal leaves are long-stalked and ovate. Even the stem leaves have few divisions compared with Garden Heliotrope. In addition, it is found exclusively in wetlands. It has been reported from 12 southern Ontario counties as well as Manitoulin District, but in recent years only from Grey, Simcoe, Hastings, Wellington and the Ottawa area. Marsh Valerian is a rare plant even in the United States. It occurs in northern and western New England, in New York State, western Michigan and northern Ohio. Southern Quebec is the only other reported location in Canada.

Local Distribution: Extremely rare, in Grey County only. Reported from Proton Township in 1935 and, in 1990 from a wetland on a sideroad west of Markdale.

Habitat: Marshy meadows, bogs and swamps.

Herbarium Specimen
(Basal leaves not present)

DESCRIPTION
Height: ± 60 cm. Fibrous rooted perennial from a stout rhizome. Stem stiff, grooved.
Leaves: Basal, elliptic, long stalked. Length ± 10 cm. Upper leaves pinnately divided, long stalked, ± 12 cm long.
Inflorescence: Crowded, short-stemmed panicle.
Flowers: Petal tube 5-lobed, 5-7 mm long. **Pink or white.** Stamens 3, protruding from petal tube.
Fruit: Achene, smooth, ovate, 3-4 mm long.
Flowering: June to July.

31. Cooper's Milk Vetch

G4 S3

***Astragalus neglectus* (Torrey & Gray) Sheldon**

Bean or Pea Family

[*Astragalus* — ancient Greek name for plant with pea-type pods, *neglectus* — overlooked (Lat.)]

The Milk Vetches (*Astragalus* spp.) are mainly western plants. Cooper's Milk Vetch appears sporadically from New York State and Virginia as far west as Saskatchewan. It appears to be somewhat dependent on the presence of limestone. It has the typical irregular flowers of the Pea Family arranged in a raceme, and the pinnately divided leaves associated with this group. There are twenty-seven species from this family listed from Grey and Bruce Counties, of which only seven are native. Of that seven two are *Astragalus* species, *A. canadensis* which is widespread and common and *A. neglectus*. Canada Milk Vetch has two-celled pods and creamy flowers. Cooper's Milk Vetch is reported from a number of counties across southern and central Ontario as far north as Manitoulin District.

Local Distribution: Uncommon overall but fairly common in the northern quarter of the Bruce Peninsula and on the Cape Croker Reserve. Also known from (a) about 8 sites in the southeastern half of Lindsay Township (b) several sites from the north part of MacGregor Provincial Park south almost to Douglas Point (c) 1 site in the North Keppel (Big Bay) area. On the Bruce Peninsula about 85 percent of the total population is east of Highway 6, as far south as the mouth of Owen Sound.

Local Habitat: Poor soil, mostly a very thin to slightly thicker layer over dolostone or limestone. Also on gravel, shale, sand and talus. In open woods of various composition and at open sites such as alvars. Fairly frequently found along roads and trails.

Herbarium Specimen

DESCRIPTION

Height: ± 25-60 cm. Stems erect, smooth and leafy.
Leaves: ± 6 cm long, pinnate, ± 15 leaflets ± 1.5 cm long.
Inflorescence: Long stalked raceme, 10-20 flowers.
Flowers: ± 12 mm long **white**.
Fruit: Erect stalkless pod ovoid, inflated 15 x 5 mm. One-celled.
Flowering: Early summer.

32. Oswego Tea, Bee Balm

G5 S3

***Monarda didyma* L.**

Mint Family

[Monardes — 16th century Spanish physician & botanist, *didymus* lobed (Lat.)]

The Mint Family is characterized by its tubular two-lipped flowers. Many species have aromatic leaves, often with culinary or herbal properties. Oswego Tea is no exception and it appears that bees among other insects are also partial to it. Hummingbirds, too, find it very attractive. All the fifteen species of *Monarda* are North American, five of which are found in the eastern half. Oswego Tea is found on moist soil in woods and thickets from Maine to Michigan, south to New Jersey and then through the mountains to northern Georgia. The early settlers found that the dried leaves of this plant made an acceptable tea, and not only that, the flowers are so striking that it became extensively cultivated. Thus, particularly in Ontario, it is not always easy to determine if a native population has been found, or if it has escaped from some old cottage garden. However, the populations in this area do seem to be native. The inflorescence is unusual in this group. Instead of the much commoner racemes or whorls found in the Mint family, the flowers, which are comparatively large, are compressed into a head-like cluster. The two-lipped crimson flowers with their protruding stamens emerge from the top like a cluster of snake's heads. It can be a breathtaking sight to be hiking through the forest along a remote river valley, and suddenly come across a patch of this striking flower.

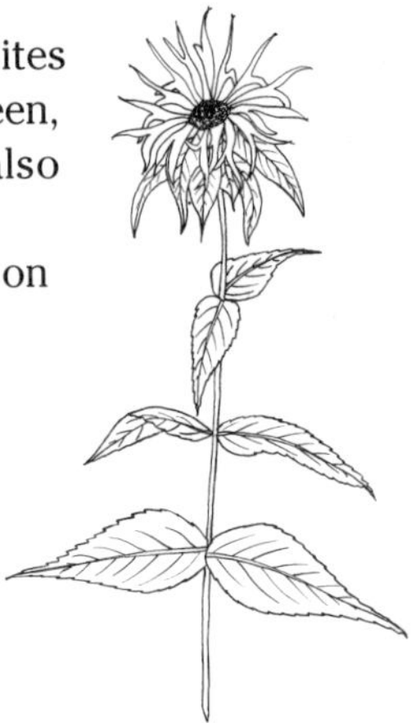

Local Distribution: Rare to very rare. Known from only 6 sites (two only 1 km apart) along the Beaver, Rocky Saugeen, Saugeen, and Sauble Rivers as well as in Greenock Swamp. It was also recorded from near Meaford before 1950.

Local Habitat: Five sites visited in the last few decades are on mesic to wet ground in mixed or, less often, deciduous forest. Some associated trees are White Cedar, Hemlock, Yellow Birch, Sugar Maple and Red Maple. All but Greenock Swamp site are in river valley bottoms. Even at that site there is a very small stream at the bottom of a slope. The Sauble River population, which is no longer in existence, was on alluvial sand but some of the other sites are rocky.

DESCRIPTION

Height: ± 50-90 cm. Stout stems, sometimes hairy.

Leaves: Ovate to ovate-lanceolate. Stalk ± 2 cm. Blade ± 9 x 4 cm. Sharp-pointed. Edge toothed.

Inflorescence: Clustered head, oldest flowers in centre.

Flowers: Length ± 4 cm. Straight upper lip. Three-lobed lower lip. Stamens protruding. Fragrant. **Crimson**.

Fruit: Four nutlets in each 5-toothed calyx.

Flowering: July to September.

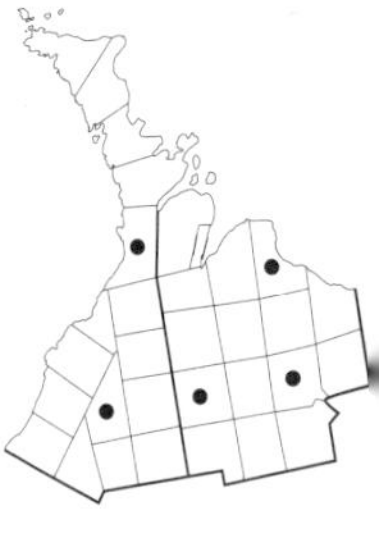

33. Narrow Leaved Mountain Mint **G5 S3**

***Pycnanthemum tenuifolium* Schrader** **Mint Family**

[*Pycnanthemum flexuosum* (Walter) B.S.P.]

[*pycnanthus* — crowded flowers (Gr.), *tenui* — slender, *folius* — leaf (Lat.)]

Narrow Leaved Mountain Mint has the standard characteristics of square stems, opposite leaves, and two-lipped flowers, that make it easy to place plants in the Mint family. However, there are two things that are unusual, one being the extremely narrow leaves and the other the crowded head-like cymes. It is also almost non-aromatic. This species is widely distributed in eastern North America from Maine to Florida and west to Minnesota, even reaching Texas. It has a spotty distribution throughout Ontario and, apart from the Ottawa area, seems to be at its northern limit here. There are a number of other closely related species in eastern North America, but there seem to be only two that penetrate as far as Ontario. *Pycnanthemum virginianum* has been found on the Bruce Peninsula, but there is no danger of confusion as the leaves are much broader and the heads more widely separated.

Local Distribution and Habitat: Extremely rare. A considerable quantity was seen near Schmidt Lake (Bruce County) in 1987, growing in small natural openings in mixed forest, although it was near a farm field. The site was calcareous and moderately moist. Also collected prior to 1979 at an unidentified location (Scoggan, 1979) in Grey County.

DESCRIPTION

Height: ± 50 cm. Slender, stiff, leafy stems arising from a creeping rhizome; many short, leafy branches.

Leaves: Length ± 3 cm, width ± 3 mm. Linear, untoothed edges.

Inflorescence: Flowers in stalked, hemispheric heads, arranged in a corymb.

Flowers: Two-lipped, **White** or purple-dotted. Corolla tube 5-8 mm long. Tubular calyx with five teeth.

Fruits: 4 nutlets in persistent calyx.

Flowering: July, August.

34. Gattinger's False Foxglove **G4 END S2**

***Agalinis gattingeri* (Small) Small** — **Figwort or Snapdragon Family**

[*Gerardia gattingeri* Small]

Most of this family have distinctly irregular flowers, often two-lipped. In *Agalinis*, the five-lobed tubular corolla has almost equal lobes. As the English name suggests they are reminiscent of the European Foxglove, but this is an exclusively American group. Gattinger's False Foxglove is a small annual plant with numerous, wiry branches and small pink flowers. This is on the extreme edge of its range here, which extends southwest to Oklahoma and Texas. It has only recently been recognized on the Bruce Peninsula and Manitoulin Island; it had previously been misidentified as Slender Agalinis. It is also known from Lambton and Brant Counties. Gattinger's False Foxglove is found in woods and hilly country as well as barrens in the main part of its range. The very similar Small Flowered Agalinis is relatively common in our area but it is a plant of moist areas.

Local Distribution: Very or extremely rare. Known for the last two decades (though misidentified for much of that time as Slender Agalinis — *Agalinis tenuifolia*) on the east side of the central Bruce Peninsula. It occurs there on the main Cape Croker prairie, where it was flourishing in 2000, and also on Barrier (Rabbit) Island. It may also have been found further north on the same side of the Peninsula, but misidentified.

Local Habitat: The Cape Croker population is on the limestone or calcareous shale of a prairie alvar, mainly in association with Grooved Yellow Flax, another provincial rarity.

Herbarium Specimen

DESCRIPTION

Height: 10-30 cm. Much branched, slender, wiry branches.

Leaves: Linear, very narrow, length ± 2 cm, width ± 1.5 mm.

Inflorescence: Single, slender-stalked flowers, terminal, or in leaf axils.

Flowers: ± 1.5 cm long **pale rose-purple**. 5-pointed calyx, tubular corolla, 5 almost equal lobes, quite "open" at the mouth, a horizontal band of long hairs at the base of the upper two corolla lobes.

Fruit: Sub-globose capsule, diameter 4 mm.

Flowering: August to early September.

35. Green Violet **G5 S2**

***Hybanthus concolor* (Forst.) Sprengel** **Violet Family**

[*Cubelium concolor* (T.F.Forst.) Raf.] [*concolor* — same colour throughout]

A Green Violet seems like a contradiction in terms. Most of us expect violets to be blue or, white or yellow. We also expect violets and pansy petals to spread out from the sepals and face up. In this flower the petals are barely longer than the sepals and the flowers hang down. However, in the number of petals and sepals and the structure of the flower and fruit, this group does belong in the Violet Family. This particular species is an outlier of a genus that is generally found in warmer climes, and our area is well north of the main part of its range, which extends from Vermont to Florida and well into the mid-west. There is only one other species in North America, out of 150 world-wide, and that is found in the southwestern states. This makes it all the more amazing that it should be discovered as far north as this. The flower is actually greenish-white, so it is distinguishable from the leaves. Nevertheless, this is not a plant that stands out. It grows in wet to mesic forests, in rich woods and ravines which are shady when it blooms. For this reason, it is probably often overlooked and may not be quite as rare as it appears. It has been reported from at least 13 counties or regional municipalities in Ontario, although in four cases not recently. This would be a plant vulnerable to habitat disturbance by logging or development. The fact that one of our populations was very substantial in numbers, even though now extirpated (see below), and the two populations were quite widely separated makes it likely that this could be rediscovered in a rich, moist deciduous forest somewhere in this area. It is certainly worth looking out for, and a challenge to spot it in the green light of the forest.

Local Distribution and Habitat: Extremely rare. Found in about 1970, northeast of Hepworth in deciduous forest. No significant attempt has been made to relocate the plants since then. A second population of approximately 200 plants was found at Kinlough (Kinloss Twp.) in 1987. This was on a loamy slope in open canopy Sugar Maple forest. It has been extirpated by gravel extraction. These populations appear to be the most northern ever reported in Canada.

DESCRIPTION

Height: 30-50 cm. Stems solitary or clustered from fibrous roots.

Leaves: ± hairy, broadly elliptic, 7-10 cm sharp pointed, tapering to slender stalks.

Flowers: ± 5 mm long **Greenish-white**. 1-3 hanging down from leaf axil. 5 linear subequal sepals nearly as long as the petals. 5 petals, subequal, one with a spur.

Fruit: 3-valved capsule, 1.5-2 cm long.

Flowering: May, June.

Other Species of Dicotyledons Listed as Rare

Historical — pre 1975

Sharp Leaved Goldenrod — G5 S3

(*Solidago arguta* Aiton) — Aster Family

This species was collected in the Durham area in 1958.

Purple Rock Cress — G5 S3? (subspecies only)

(*Arabis divaricarpa* Nels. var. *dacotica*) — Mustard Family

Collected in 1932 on the Cape Croker Indian Reserve and identified as this variety. The differentiation of this specimen from the other variety of *A. divaricarpa*, which is common here, is perhaps questionable.

Shrubby St. Johnswort — G5 S2

(*Hypericum prolificum* L.) — St. Johnswort Family

Collected in Grey County (unspecified location) in 1841 and at Clarksburg in 1943. It has not been recorded for the last 25 years.

Long Stemmed Waterwort — G5 S3

(*Elatine triandra* Schkuhr.) — Waterwort Family

There is a historical report from near Stokes Bay, but no specimen was collected and it may have been misidentified.

Clustered Broomrape — G4 S1

(*Orobanche fasciculata* Nutt.) — Broomrape Family

There is a historical report of a photograph taken at Dorcas Bay on the Bruce Peninsula, but there is some doubt as to the accuracy of this report.

Bluets — G5 SH

(*Hedyotis caerulea* (L.) Hooker) — Bedstraw Family

Collected at Whitechurch at the southern edge of Bruce County in 1970.
There have been no subsequent reports.

Not native to this area, "not definitely S1-S3," or unconfirmed:

Rayless Aster — G5 S3?

(*Aster brachyactis* S.F. Blake) — Aster Family

This species is considered to be an introduction, possibly associated with salted roadsides. It is very rare here but has been found on the Bruce Peninsula and in other parts of both counties. It is not a rare species in its native habitat in its main range in central and western North America. Also occurs in coastal salt marshes in southern James Bay.

Provancher's Fleabane **G1G2 SU** (subspecies only)

Aster Family

(*Erigeron philadelphicus* L. ssp. *provancheri* (Victorin & Rousseau) Morton)
There is controversy as to whether the small plants with white flowers growing along the rocky shorelines in the north of the Bruce Peninsula mainland and on some of the islands are actually the same as the Provancher's Fleabane found along the St. Lawrence in Quebec. However, these plants are probably rare in Ontario, and they have been included in rare plant lists in the past.
More detailed information about the two species above can be found in **The Asters, Goldenrods and Fleabanes of Bruce and Grey Counties**.

Bugseed **G1G3 S1S3**

(*Corispermum* sp.) Goosefoot Family
Recent changes in nomenclature have made it difficult to be sure which of four possible species, now considered to be native North American, is found here. There is no clear description available which would help us separate them. The species we have is certainly rare. It was seen by Joe Johnson in the 1980s and Anwar Maun in the late 1990s at three slightly disturbed locations along the west side of Bruce County in the Sauble Beach area and at Inverhuron. Two of the three locations are at the Lake Huron shore, found on dry sand in the open and semi-open. It seems that even if all four species were treated as one it would still be provincially rare and rated S3. At this time, there are no illustrations or material available which would assist the amateur in clear separation of the species or their identification.

Whorled Milkweed **G5 S2**

(*Asclepias verticillata* L.) Milkweed Family
Found once in an old field near Flesherton in 1983. Probably adventive.

Meadow or Prairie Sundrops **G5 S2**

(*Oenothera pilosella* Raf.) Evening Primrose Family
Found in Holland Township in 1999. Probably a garden escape and most likely not deliberately planted in that location. An earlier report from the Bruce Peninsula is most likely an error.

Golden Seal **G4 THR S2 THR**

(*Hydrastis canadensis* **L.)** Buttercup Family
A substantial population of this plant has existed for several decades in one woodlot between Durham and Dornoch in Grey County. It was originally brought from a more southern location in Ontario and planted. Since then it has reproduced and spread around the woodlot.

(Northern) Fox Grape **G5 S1**

(*Vitis labrusca* L.) Grape Family
Escaped from cultivation in this area. Known from a few locations in central Bruce County.

FLOWERING PLANTS
Monocotyledons

Plants in this group have the following distinctive features in common:
1. They produce a single leaf when the seed germinates. (All other flowering plants produce two seed leaves).
2. Their leaves have parallel veins and are often long and narrow.
3. The parts of the flower are in threes or multiples of three.

Several families of moncotyledons are represented among our rare species.

a. **Water Plantain Family** (page 105)
   Rooted aquatic plants of shallow water.
b. **Arum Family** (page 106)
   Often known as Arum Lilies, typified by the common Jack-in-the-Pulpit. "Jack" is the erect, fleshy spike (spadix) covered in tiny flowers, and the "pulpit" is the large leafy bract (spathe) surrounding it. Usually, the tiny flowers are single sexed — those with stamens (male) at the top of the spike, and those with pistils (female) at the bottom. Many species are pollinated by flies which crawl into the bract, attracted by the disgusting (to us) smell of rotting meat emitted by the plant. The leaves have a main vein and each half is parallel veined. Mostly associated with damp areas, or aquatic habitats.
c. **Sedge Family** (page 107)
   This very large family includes Cotton Grass and some types of rush as well as sedges, with over 4000 species worldwide. In Bruce and Grey alone there are 104 species of sedge and 34 species from other groups. Of these only 5 species are considered provincially rare. Members of this family tend to grow in wet areas and are specially vulnerable to changes in drainage patterns. Most can be recognized by their solid stems (except for some rushes) and usually grass-like leaves. Like the grasses, the leaves consist of a closed basal sheath, usually expanding into a blade.The tiny flowers are either pistillate or staminate. They have no petals or sepals but are enclosed in a single scale. They are grouped in dull-coloured spikes. Most species have triangular stems. "Sedges have edges!"
d. **Iris Family** (page 112)
   This is a comparatively small family, only 1500 species world-wide. They are closely related to lilies. The flowers are colourful, often blue. Each flower has 6 equal coloured segments (tepals) united at the base into a tube. The ovary is below the tepals (inferior). In lilies the ovary is above the tepals (superior) and the tepals are most often separate. The leaves are strap-like.
e. **Orchid Family** (page 113)
   One of the most complex families with over 15 000 species world-wide. This is considered to be the most evolutionarily advanced family. They are extremely habitat specific and dependent on insects for pollination. These often colourful flowers have 3 sepals, often coloured, alternating with 3 petals, two similar and the third petal enlarged to form a lip or pouch.
f. **Grass Family** for details see page 120.
g. **Pondweed Family** see page 127.
h. **Bur Reed Family** see page 128.

36. Crested Arrowhead **G4? S3**

***Sagittaria cristata* Engelman** **Water Plantain Family**

[*Sagittaria graminea* Michaux var. *cristata* (Engelm.) Bogin]

[*Sagitta* — arrow; *cristatus* — crested (Lat.)]

The Arrowheads are a group of aquatic plants which root in shallow water and produce rhizomes, tubers, or corms. Some species are known as Duck Potatoes. When flowering, the tall stalks with whorls of three-petalled flowers are easy to recognize. Many of them, such as *S. cuneata* (Northern Arrowhead) and *S. latifolia* (Common Arrowhead), have arrow shaped leaves, but this rare species is distinguished by its very narrow grass-like leaves, often without any blade. Crested Arrowhead (*Sagittaria cristata)* is an uncommon relation of the much more widespread eastern North American plant Grass Leaved Arrowhead (*S. graminea*) which is very similar, but it has not been found in our area. They are separated by the differences in the structure of the one-seeded achenes and the stamens are also different. Sessile Fruited Arrowhead (*S. rigida*) has also been reported from the peninsula in the past. It does not always have arrow shaped leaves but they tend to be wider than in this species, and the fruits are stalkless. Crested Arrowhead is only found in Michigan, Minnesota, and Wisconsin with outlying populations in Missouri and Nebraska as well as Ontario. In this province, besides Bruce County it occurs in Manitoulin District, and has been reported from eight other counties or districts. In many books it will be treated merely as a variety of Grass Leaved Arrowhead.

Local Distribution: Very rare in our area, known only along the west shore of the Bruce Peninsula; seen at several locations from near Tobermory to near Chiefs Point. Reported from the 1920's until at least 1993. Not recorded from Grey County. Grass Leaved Arrowhead is not provincially rare, and has not been reported from Bruce or Grey Counties.

Local Habitat: Shallow water along the Lake Huron shoreline. In shallow, protected bays; the only population seen by Joe Johnson was at a creek mouth.

DESCRIPTION

Flowering Stems: ± 20 cm.

Leaves: Lance-linear blades, some stalked. 3-5 parallel veins, 8 x 1 cm.

Inflorescence: 2-3 whorls of 1-2 flowers. Upper flowers staminate, lower pistillate.

Flowers: Diameter up to 1 cm. Petals 3, **white**. Stamen — anther shorter than filament. Sepals 3, recurved to spreading.

Fruit: Fruiting cluster on a stalk, ± 1.5 cm diameter. Achenes, ± 2.6 x 1.6 mm. Wing with wavy edge on back. Straight wing below. Each side with 2 wavy crests. Clustered on a stalk. (*S. graminea* slightly winged on the margins, ribbed on the sides — not crested.)

Flowering: July to September.

37. Arrow Arum G5 S2

***Peltandra virginica* (L.) Schott. & Endl.** **Arum Family**

[*Peltandra* — shield shaped — male flowers (Gr.); *virginica* — from Virginia]

Arrow Arum is a plant of swamps and shallow waters, with large, bright green, arrow-shaped leaves. The bract (spathe) is green and long and narrow. It curves back at the tip, and covers the spike completely but opens halfway to expose the staminate flowers and allow access to the pollinating insect. The berries are brownish green. The lack of distinguishing colour and the watery habitat make this plant hard to spot, but the large arrow-shaped leaves are distinctive. The leaves of most species of Arrowhead, which are also arrow-shaped, have the veins fanning out from the top of the stalk. In Arrow Arum they branch from each side of the midrib. Arrow Arum is very much on the northern edge of its range here. It occurs in the States from Maine, west to Michigan and as far south as Florida and Texas. In Canada, it is found in Quebec and Ontario.

Local Distribution and Habitat: Very rare. Known from only two locations in Bruce County, one in a small shallow bay or marshy shore at one of the Fishing Islands off Oliphant (1984), the other at Cunningham Lake in Greenock Swamp in 1987. At the latter site, there were probably at least 1000 plants. They grew around the margin of the lake (a) in open quaking vegetation (b) at the water's edge (c) on islets near the shore and (d) a few plants in adjacent deciduous swamp habitat.

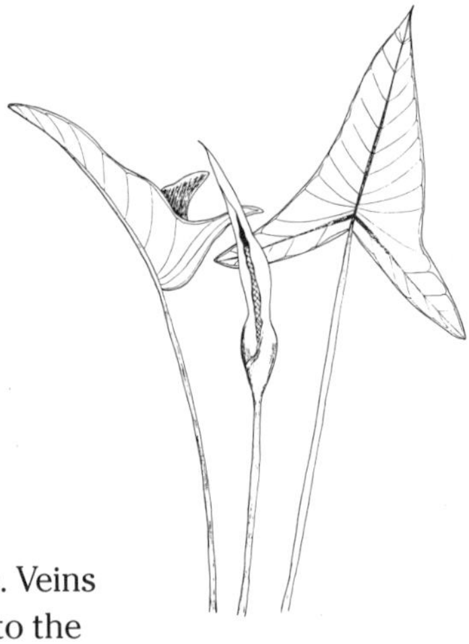

DESCRIPTION

Leaves: Arrow shaped. 10-25 cm long, 7.5-15 cm wide. Veins parallel in each half of the leaf from a central midrib to the border.

Inflorescence: Stem 20-40 cm. Slim bract 10-20 cm green with a pale margin.

Pistillate flowers ± lower half of spike. Staminate flowers ± upper half of spike.

Fruit: Head of brown berries, 1-3 seeded, in "jelly."

Flowering: Uncertain.

38. Appalachian Sedge G4 S2S3

Carex appalachica **Webber & Ball** **Sedge Family**

[*Carex radiata* Schk. ex Willd. — this name is now used for a closely allied species]
[*kerein* — to cut (Gr.) sharp edged leaves)

This small, slender sedge is difficult to distinguish from two much commoner species — Star Sedge *C. rosea* and Turf-like Sedge *C. radiata*. It is slightly more delicate than either of them. The flowers are in small few-flowered tufts on wiry, flexible stalks. This is another plant which is on the edge of its range here. Although it is a plant of cool temperate forests, it only extends from Maine into Ontario and Quebec, as well as to New Jersey and Pennsylvania. Beyond that, it is found in the Appalachian Mountains as far as North Carolina and Tennessee.

Local Distribution: Very rare. Definitely known from only one location above the Niagara Escarpment between North Keppel (Big Bay) and Lake Charles. It was also reported from a site above the escarpment a few kilometres away. Both reports date from the 1980s. These are the most northerly locations known in Ontario.

Local Habitat: In both cases reported from upland calcareous, Sugar Maple forest.

Note: There are two local species closely allied to Appalachian Sedge so it could have been overlooked at one or more additional Grey-Bruce localities; nevertheless it is probably genuinely very rare.

Herbarium Specimen

DESCRIPTION

Height: 20-30 cm. Dense low tufts.

Leaves: ± 1 mm wide

Inflorescence: 2-3 very small rounded clusters on each stem. 4-7 flowers. Upper flowers staminate, lower pistillate.

Flowers: Enclosing sac ± 2.5 mm

Fruit: Achene, ovate, ± 2 mm long.

Flowering: Uncertain. Probably late May-June.

39. Schweinitz's Sedge **G3 S3**

Carex schweinitzii **Dewey** **Sedge Family**

This sedge is confined to temperate northeastern North America from Vermont to northern Michigan, south to New Jersey, North Carolina and Missouri, and is rare throughout its range. In Ontario it has not been found east of Northumberland County, being mostly in the southwestern counties. Apart from a historical record from Manitoulin District, the Bruce Peninsula is about its northern limit. The restrictions of habitat (see below) will be one clue that this particular sedge has been found. There is usually one slender, terminal staminate spike, with two to five pistillate spikes lower down the flowering stem. These spikes are quite large, from 3-7 cm long. They are cylindrical and ascending, the flowers are not very densely packed. Generally, sedges found in this area have shorter, denser, spikes or they hang down. Nevertheless, this species is quite similar to a few of our common medium to large species. Expert assistance may be needed to help in its identification.

Local Distribution: Uncommon in Grey County. Rare in Bruce County. Known as far north as Niagara Escarpment sites between Oxenden and North Keppel (Big Bay), on the Bruce Peninsula in Grey County.

Local Habitat: This is a sedge of seepage areas and springs, as well as borders of cold rivers and creeks. It is mainly found at sites somewhat thinly shaded by trees, most often mixed stands or even coniferous stands rather than deciduous.

DESCRIPTION

Height: 30-70cm. Tufted.

Leaves: Lower leaves reduced to scales. Remaining leaves above base of plant.

4-10 mm wide, roughish.

Inflorescence: Upper spike staminate. Lower spikes (2-5) pistillate, 3-7 cm long ascending, flowers loosely packed. Bracts at base of spikes longer than spikes.

Flowers: Enclosing sac light green ± 2.5 mm with a 2-toothed beak.

Fruit: Achene, 3-sided, sides concave. Style persistent, flexuose.

Flowering: June to July.

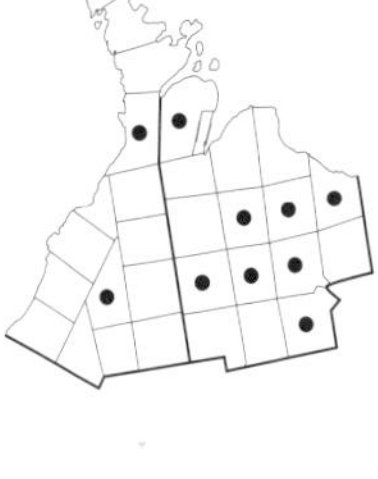

40. Rigid Sedge (Common Stiff Sedge)

G4G5 S3

***Carex tetanica* Schk. ex Willd.**

Sedge Family

[*tetanica* — rigid]

This sedge is more widely distributed than the other two species listed here, being found from New England through the northern states and Manitoba as far west as Alberta and Nebraska. It has also been reported from a number of counties etc. ranging from the Ottawa area in the east to Essex in the southwest, and north to Cochrane District.

Local Distribution and Habitat: Very rare. Known only from two locations in central and southern Bruce County, dating back to 1987. One site was in Greenock Swamp, in mixed swamp forest, where there were a few plants. The second was adjacent to the Saugeen River, in an open flood plain, where there were 50 to 100 plants.

Note: Both populations were found by an exceptionally keen-eyed professional botanist, so it may well be that this species is easily overlooked. It is, nonetheless, probably very rare.

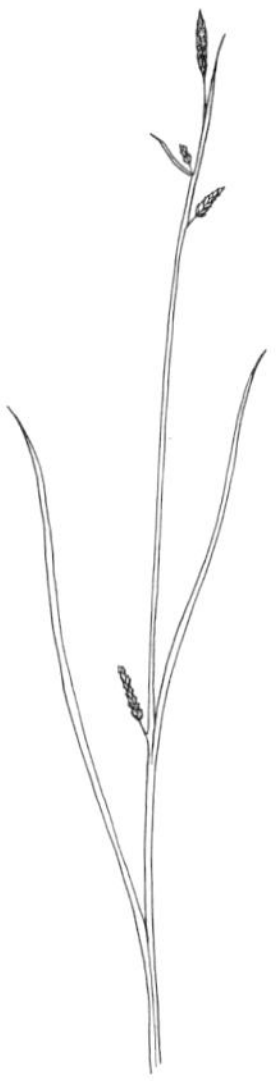

DESCRIPTION

Height: 12-25 cm. Dense tufts.

Leaves: Thin, green, soft, almost translucent. 1.5-4 mm wide, older leaves rolled back.

Inflorescence: Long-stalked terminal staminate spike. 1-3 pistillate spikes. Pistillate scales broad, egg-shaped, purplish-brown margins, green centre, smooth.

Flower: Enclosing sac, 2 lateral ribs, broadly tapered or rounded.

Fruit: Achene, broad oval, wider at top, concave sides.

Flowering: Uncertain.

Note: This species could be mistaken for the common *Carex aurea* (Golden Sedge), although the habitat is different.

41. Beaked Spikerush (Wicket Spikerush)

G5 S3

***Eleocharis rostellata* (Torrey) Torrey** **Sedge Family**

[*eleo* — marsh, *charis* — grace (Gr.);
rostrum — beak (Latin) referring to the "beaked" achene]

The spikerushes are a small group, with only 150 species world-wide. There are 10 species in this area, of which this is the only one designated provincially rare. When Robert Brown coined the genus name in 1810 he must have been thinking of how these small plants "grace" marshes, often in what we would describe as "mucky" conditions. The miniscule flowers are grouped into small terminal spikelets. Beaked Spikerush is present in both eastern and western North America, with a gap in between. The range barely extends into Canada, but continues south as far as tropical America. It is very much on the edge of its range here.

Local Distribution: Overall this species must be designated uncommon in this area. Nevertheless, it is more common than elsewhere in Ontario, and probably much easier to find. In Bruce County it is known from more than 20 locations, from the northern side of Bruce Peninsula National Park south to the Kinloss-Teeswater-Formosa area. It is less often found in Grey County, although there are a few records, in the northwestern quarter of the county as far south as near Holland Centre. When present, it is often abundant, and sometimes is one of the dominant plants in that location.

Local Habitat: Usually in marly fens i.e. calcium-rich open wetlands, mostly quaking or near or on the west shore of Bruce County. Many of these fens, especially inland, have quaking sedge mats with extremely watery muck below. Sometimes these mats are found near or at the edge of a lake. Occasionally found in other open marly sites, often on relatively barren, saturated muck. Definitely a calciphile. In marly wetlands, look for a rather coarse spikerush with long, hard, sterile stems that arch over and sometimes root at the tips (the origin of the alternative English name "wicket" spikerush).

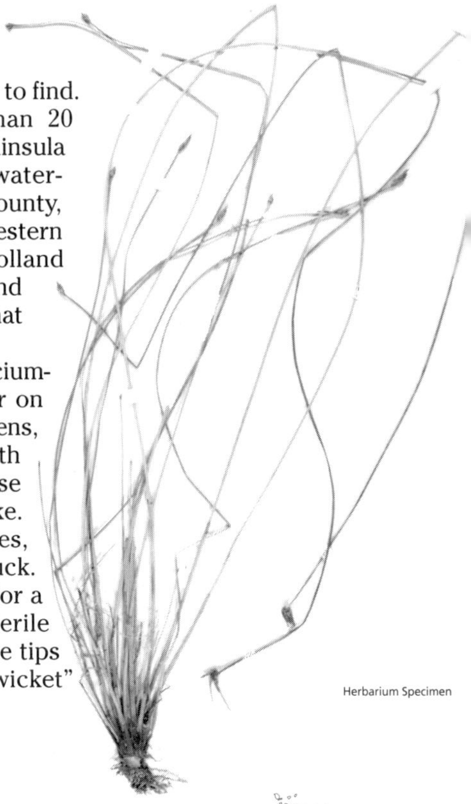

Herbarium Specimen

DESCRIPTION
Height: 10-60 cm. Often in dense colonies.
Inflorescence: Stems 1-2 mm wide with a single, terminal spikelet.
Spikelets: Spindle-shaped. 8-10 mm long. 10-20 flowers.
Flowers: Bisexual, 3 stamens and a 2-3 parted style.
Fruit: Achene ± 2 mm long. Light greenish-brown. Beak extending smoothly from the achene with no ridge.
Flowering: Early summer (and late May?)
Fruiting: Summer, early autumn.

42. Low Nut Rush

G5 S3

***Scleria verticillata* Muhlenb. ex Willd.**

Sedge Family

[*scleros* — hard (Gr.)]

This is another rather small group of the Sedge family with only 200 species worldwide. The genus name refers to the characteristic hard, white (crustaceous) achenes found in this group. Again, Low Nut Rush is very much on the northern edge of its range here. It is widely distributed in eastern North America from New England to Minnesota and south into the Caribbean and Mexico. The spikelets in this group are pistillate or staminate. Female spikelets have one flower, male spikelets several. They are found in tiny bunches at intervals on the thin, wiry flowering stem, which is three-angled. The white "seeds" provide the most definite means of identification. They are tiny, but with a lens it can be seen that they are shaped like a child's top and ridged. In this species they contrast with the blackish disk below — a good way of identifying Nut Rushes in the field. This is the only species of *Scleria* found in this area so there is little possibility of confusion. Another helpful feature is that it can often be identified, when flowering and still inconspicuous, by the pleasant hay-like odor produced when it is walked on.

Local Distribution: Uncommon. Known only from the Lake Huron shoreline in Bruce County, occurring sporadically from between Inverhuron and Douglas Point north to Dorcas Bay. In the Oliphant area it can be described as "common" along several kilometres of shoreline.

Local Habitat: Nearly always found in the open, mostly either on marly sand of moist meadows, or in drier than average marly fens. Occasionally on marly, gravelly or sandy shores. Often found on cottage lawns in the Oliphant area and elswhere in the southwest part of the Bruce Peninsula. It has been found in marly wet meadows that are not sandy and once on a low alvar. Always less that 1.3 km from Lake Huron (at average lake level), often in view of the water.

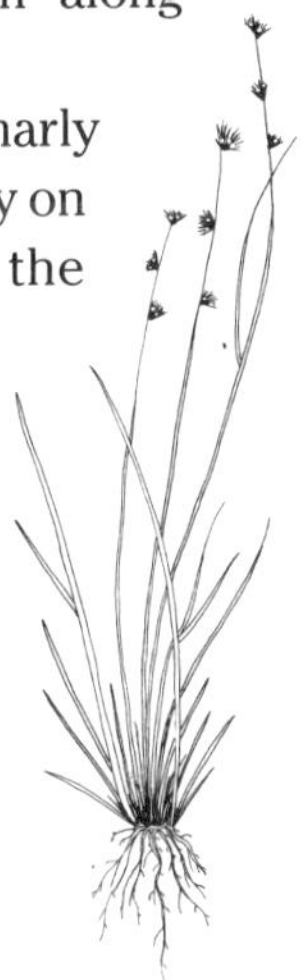

DESCRIPTION

Height: 10-15 cm or more. Often in dense patches.

Leaves: Sheaths slightly hairy. Blade 1 mm wide.

Inflorescence: An interrupted spike with sessile spikelets.

Spikelets: Staminate with several flowers. Pistillate with one flower.

Fruit: Achene 1 mm long. "Top" shaped, round with a sharp point. Transversely ridged. **White** on a black disk.

Flowering: Late July, August.

Fruiting: August to October. Annual with overwintering seeds.

43. Dwarf Lake Iris

G3 S3

***Iris lacustris* Nutt.**

Iris Family

[*Iris* — rainbow (Gr.); *lacus* — lake (L.)]

Anyone who has ever seen a cultivated iris will have no difficulty recognizing this flower, and it is certainly dwarf compared with a Dutch Iris or a Bearded Iris. This is an endemic species found only around the upper three Great Lakes, almost restricted to the northern parts of the main body of Lake Huron and Lake Michigan. Most of Canada's total population is found in Bruce County. The only other dwarf iris in North America is in the east — found from Ohio south, so there is no danger of confusion. In our area, large colonies often form which do not bloom every year. The production of fruit is uncommon and it mostly reproduces by means of creeping rhizomes. Stands of it here can be very extensive. Ironically, most people who live on the west side of the Bruce Peninsula would probably regard it as common and quite expendable, whereas in global terms it is a rare plant, and is locally very vulnerable to cottage development.

Local Distribution: Mainly on the Bruce Peninsula (including the Tobermory Islands) and continuing along the Lake Huron shore in Bruce County as far south as Inverhuron. On the peninsula it can be considered fairly common and is often locally abundant. It is mainly found within 1 km of the west shore, but in the northern part of the Bruce Peninsula very extensive stands occur up to 7 km inland. These are found in Jack Pine and Red Pine areas. It is absent east of Highway 6 except in a few places in and adjacent to St. Edmund's Township. Everywhere further south than Wiarton it is only found within half a kilometre of the Lake Huron shore. It is common near the shore in MacGregor Point Provincial Park, and towards the Lake Huron shore of Bruce Peninsula National Park (west or southwest side).

Local Habitat: Wet to dry ground with poor soil. On sand or sandy loam, dolostone bedrock or gravel (less often). Usually in woodlands, mostly conifer dominated and often quite open, under white cedar (including colonies in cedar swamps), white birch, balsam fir, red pine, jack pine, tamarack, white or black spruce. Sometimes in the open in alvars, or fens, or on shores. Usually found under calcareous conditions, but sometimes associated with Trailing Arbutus which requires an acid site.

DESCRIPTION

Height: 15 cm. Forming large clones.

Leaves: Light green, slightly waxy. Broadly linear, curved, ± 1 cm wide. Change to distinctive white when dead in autumn.

Inflorescence: Flower solitary.

Flower: 5-6 cm wide. Outer 3 petals reflexed, notched, with **whitish-yellow beard** in darker centre. Inner 3 petals more erect — **blue**

Fruit: Capsule, ovoid, 2 cm. Dark brown seeds.

Flowering: May to mid-June, rarely earlier or later.

44. Putty Root, Adam & Eve G5 S2

***Aplectrum hyemale* (Muhl. ex Willd.) Nutt.** **Orchid Family**

[*a* — without; *plectrum* — spur (Gr.); *hyemalis* — wintry (L.)]

The Latin name *hyemalis* refers to the single overwintering leaf, with its distinctive pale grayish-green colour and white veins. It often disappears before the plant blooms in the spring. This may seem strange but it is an adaptation to growth on the floor of a temperate deciduous forest where light levels and moisture are very low in the summer. During the fall and early spring the new leaf is able to store food in the corm. Although it does not bloom every year, the plant produces a new corm which remains attached to the old one with a thin rhizome. Unfortunately, the sticky quality of their corms made them attractive to early settlers in the eastern United States, who used them for mending broken china — hence the name "Putty Root." The corms are also edible when cooked. The pairs of corms are the origin of the other common name "Adam & Eve." This led to numerous superstitions about this plant and attempts to use it for medicinal purposes. It probably worked as a placebo but not much else!

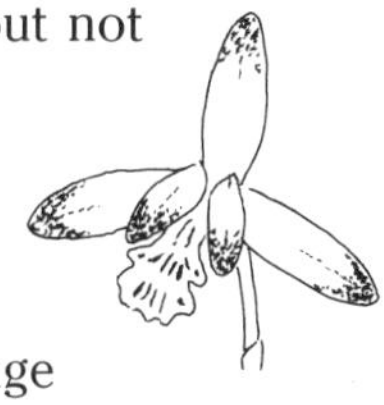

The brownish flowers make this a hard plant to spot against the lush vegetation and dead leaves of the forest floor. It is most easily found by looking for the rather striking leaf in autumn or very early spring. Putty Root is an eastern North American plant, the range extending from southern Quebec to Minnesota and south to Georgia. It has apparently disappeared from a number of the twenty-one counties etc. from which it has been reported in Ontario. It appears never to have been found in Grey County even though it does occur to this day in Simcoe. The main reasons for its apparent decrease in Ontario are logging and clearance for farming.

Local Distribution: Very rare. Putty Root has been reported from only three locations in the central Bruce Peninsula, including one on the Cape Croker Reserve.

Local Habitat: Upland deciduous forest floor.

DESCRIPTION

Height: ± 20 cm.

Leaf: Single, basal, elliptic 10 x 4 cm, stalk 2-3 cm. Dark green, whitish veins, gradually turning brown in late spring.

Flowers: 5-10 in a loose raceme, stalk with a few sheathing bracts. Sepals and lateral petals ± 1 cm, oblanceolate, **purplish, tipped with brown**. Lip obovate, ± 1 cm, 3-lobed. Central lobe white, marked with purple. No spur.

Fruit: Drooping elliptical capsule, 2 cm long.

Flowering: June.

45. Autumn Coralroot

G5 S2

***Corallorhiza odontorhiza* (Willd.) Nutt.**

Orchid Family

[*korallion* — *rhiza* — coral root (Gr.); *odonto* — tooth (Gr.)]

One of the most striking features of the Coralroots is their root system which, as their name implies, is fragile, multi-branched and looks like a piece of coral. This root system has the function of extracting nutrients from the soil as these plants are indirectly saprophytic and do not rely on photosynthesis for their nourishment as other green plants do. The specific name refers to the swollen base of the stem which looks like a tooth. The autumn flowering habit of this small plant makes it difficult to spot and, like many orchids, it does not bloom every year. It is very much on the northern edge of its range here which extends from temperate eastern North America to Mexico. Most reports in Ontario are from the Carolinian zone of the southwest, where it is associated with oak and pine on sandy soils. P.M.Catling says that it is becoming commoner in the northern part of its range. (Oldham, 1999).

Local Distribution: Extremely rare. Reported from only one pine plantation on the Bruce Peninsula. It was possibly introduced with pine seedlings from the provincial tree nursery at St. Williams. This is apparently the most northerly record of this species and is well beyond its normal range.

Local Habitat: Pine plantation on sand.

DESCRIPTION

Height: 6-16 cm.

Stem: Base swollen, purplish brownish, with a few sheathing scales.

Leaves: None.

Flowers: Lax raceme, 5-15 small flowers. Sepals and lateral petals lanceolate, **yellowish-purplish-green**. Lip obovate, broad, deflexed, **whitish, purple spotted**. Spur small at the top of the ovary.

Fruit: Capsule, ovoid, drooping, 5 x 4 mm.

Flowering: August to October or early November.

46. Ram's Head Lady's Slipper **G3 S3**

Cypripedium arietinum **R.Brown** **Orchid Family**

[*Cypris* — Venus, *pedilon* —shoe (Gr.); *aries* — ram (Lat.)]

The name of this species refers to the fancied likeness of this flower to the head of a bearded, charging ram. Ram's Head Lady's Slipper belongs to a group of orchids which have the lower petal in the form of a slipper. This is the smallest and least conspicuous of our Lady's Slippers as well as being the rarest except for the Small White Lady's Slipper. The "slipper" is a small, downward-pointing cone, the top covered with downy white hairs and the basic white veined with pinkish red. This species is quite widespread in the cooler part of the temperate forest region from Massachusets to Minnesota and south central Manitoba, but no further south than New York State. In Ontario it is known from Thunder Bay District in the north and is recorded from 28 counties etc. It is probably more common on the Bruce Peninsula and Manitoulin Island than anywhere else in Ontario.

Local Distribution: Rare overall. Merely uncommon on the Bruce Peninsula. Very rare further south in Bruce County where it appears to be restricted to the Lake Huron shore area. It has been recorded from the Tobermory Islands south to the vicinity of Douglas Point. The major part of the total population is within 2 km of the west shore of Bruce County. It could be considered fairly common in the section of St. Edmunds Township west of Highway 6.

Local Habitat: Dry places. Most often on bedrock, often with a thin film of black sandy soil. Often near alvars. Less often on sand. Mostly under very open to semi-open conifer woods — sometimes merely scrubby. Occasionally in open areas with a few scattered trees. Mostly associated with pines and cedar. It exhibits a strong affinity for natural Jack Pine stands which are almost restricted to the upper peninsula. Generally a calciphile, although it is also known from an acid Black Spruce swamp in Keppel Township.

Note: Extensive forest fires on the Bruce Peninsula about one century ago probably created habitat for Ram's Head Lady's Slipper and other rare plants such as Hill's Thistle, causing them to become commoner than in earlier times. Regenerating forest would reverse that effect.

DESCRIPTION

Height: 10-24 cm. Stem slender, somewhat hairy. 2-3 sheathing scales at base.

Leaves; 3-5 spiralling around stem above scales. Blue-green, elliptical, 5-10 x 1-1.5 cm.

Flowers: Solitary. Stalk 5-15 cm. Sepals and lateral petals linear, **greenish-brown**. Lateral sepals not joined. Slipper 1.5-2.5 x 1-2 cm, conical pouch. **White, veined with dark red or madder**. Mouth covered with white hairs. Opening circular. No spur.

Fruit: Capsule ± 3 cm long, obliquely ellipsoid, sub-erect.

Flowering: May 20 to June 20, occasionally earlier or later.

47. Tubercled Orchid **G4 S4**

Platanthera flava* (L.) Lindley var. *herbiola (R. Br.) Ames & Correll **Orchid Family**
[*Habenaria flava* (L.) R.Br.]
[*platys* — wide; *anthera* —
anther (pollen-bearing part of stamen) (Gr.) *flavus* — yellow (L.)]

The *Platantheras* are a large group of orchids, also known as the Rein Orchids which have undergone a number of name changes during the history of modern Botany. A number of them, like Tubercled Orchid have greenish-yellow flowers. As its English name suggests, this one is distinguished by having a small lump in the middle of the lip. This variety of this eastern North American species is found from Nova Scotia as far west as Minnesota and south, but a different variety is found from Florida to Texas. This is, therefore, another plant which is on the northern edge of its range here. Tubercled Orchid is always found in bogs, or swampy ground, or river flood plains. It could be confused with the common Tall Northern Green Orchid but the tubercle on the lip and the shiny leaves distinguish the Tubercled Orchid. The Long Bracted Green Orchid might cause confusion, but this has green bracts under the flowers which are three times longer than the flowers. In the Tubercled Orchid they are much shorter.

Local Distribution: Rare. Known only from Bruce County at three spots in a small section of the extensive Greenock Swamp, and at one place on the central Bruce Peninsula.

Local Habitat: Deciduous swamps, including a swamp fringe on the edge of an inland lake. Where present, it may develop substantial colonies.

DESCRIPTION
Height: 13-35 cm. Stem erect, leafy.
Leaves: 1-5, mid-green, shiny, lanceolate, sheathing stem. 5-20 x 1-5 cm, upper ones becoming small bracts.
Inflorescence: Spike 10-40 flowers.
Lower flower bracts ± 3 cm, smaller above.
Flowers: Sessile, **yellow-green.** Dorsal sepal ovate, obtuse, concave 4 x 2.5 mm. Lateral sepals ovate, oblique 4 x 2.5 mm. Lateral petals ovate, obtuse, with dorsal sepal forming a hood. Lip oblong, rounded, ± 5 x 3 mm. Triangular lobe on each side. Tubercle (lump) in centre near the base. Spur slender, club-shaped. 5-7 mm.
Fruit: Capsule ellipsoid, semi-erect, 8 x 4 mm
Flowering: Mid-June through July. Flowers often persisting on the developing capsules for several weeks.

48. Eastern Prairie White Fringed Orchid

G2 SC S2

Platanthera leucophaea **(Nutt.) Lindley**

Orchid Family

[*Habenaria leucophaea* (Nutt.) A.Gray]
[*leucon* white, *phoios* — dusky (Gr.) i.e. off-white or creamy]

This attractive plant with its spires of creamy-white flowers is a native of wet prairies or bogs and marshes. Intensive farming and the drainage associated with it have destroyed its habitat in many places, so that today more of it is to be found in sedgy fens and meadows. When not flowering, it is difficult to see among the grasses and members of the sedge family with which it is associated. Examples are Hair Grass (*Deschampsia caespitosa*), Cord Grass (*Spartina pectinata*) and Twig Rush (*Cladium mariscoides*). It seems to have disappeared from many of the counties etc. where it was once found, but it has been reported recently in Essex and Lambton, Leeds and the Ottawa area, and Simcoe and York as well as this area. Its range extends from New England into Illinois, Wisconsin, and southeast Iowa.

Local Distribution: Very rare. Found around one lake on the northern Bruce Peninsula and at one location in central Grey County but not in southern Bruce County.

Local Habitat: Two types (a) lake margins, including adjacent meadows (b) one spot in a slightly marly fen.

DESCRIPTION
Height: 10-60 cm, mostly less than 40 cm. Stem angled, leafy.
Leaves: Elliptic to lanceolate, keeled, pointed. Lower leaves 8-20 x 2-5 cm. Upper leaves progressively smaller.
Inflorescence: Raceme, 15-40 flowers. Flower bracts 1-4 cm long, lanceolate, pointed.
Flowers: Cream coloured, fragrant. Sepals 8-15 x 5-8 mm, green or greenish-white, ovate, concave. Lateral petals wedge to fan shaped, shaggy toothed. Lip 1.5-3 cm long, deeply 3-lobed, lobes fringed. Spur slender, club-shaped 2-5 cm long.
Fruit: Capsule, ellipsoid, 18 x 9 mm.
Flowering: July to early August.

49. Large Round Leaved Orchid or Dinner Plate Orchid

G4 S2

***Platanthera macrophylla* (Pursh) Lindl.**

Orchid Family

[*P. orbiculata* var. *macrophylla* (Goldie) Luer]
[*orbis* — circle (Gr.); *macros* — large, *phyllum* — leaf (Gr.)]

There should be no difficulty distinguishing the Large Round Leaved Orchid from all species except the very closely allied Round Leaved Orchid. The alternative English name — Dinner Plate Orchid — is very appropriate as the two basal leaves can be up to 15 cm across and 20 cm long. In the ordinary Round Leaved Orchid (*P. orbiculata*), which is commoner, they are half that size. The spur in *P. macrophylla* is also much larger, often twice the size of the smaller species. This sounds simple, but may be harder to determine in the field as both species are variable. The Large Round Leaved Orchid is confined to eastern North America from Newfoundland south to North Carolina and west to northern Wisconsin. The Round Leaved Orchid is much more widespread being found further north in Labrador and also on the west coast from Alaska to Oregon. The Large Round Leaved Orchid is rare in Ontario. It has been recorded at some time from ten other counties etc., but has not been reported recently from a number of them.

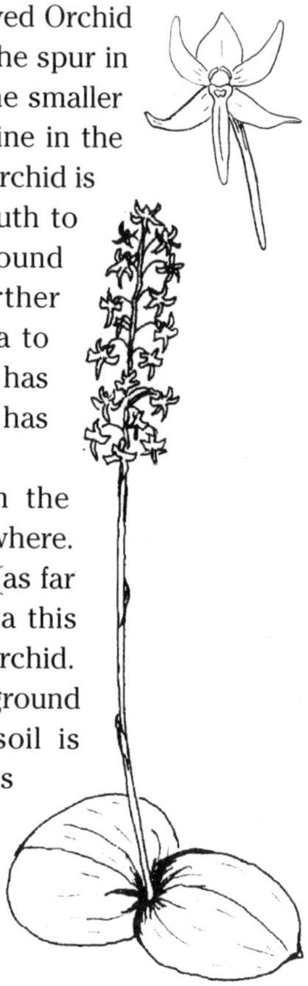

Local Distribution: Rare. Largely at various sites in the Greenock Swamp, also known at three general locations elsewhere. On the Bruce Peninsula known only from near Howdenvale (as far back as 1926) but now almost extirpated. Off the peninsula this species is considerably more prevalent than Round Leaved Orchid.

Habitat: Deciduous to mixed forest on slightly higher ground adjacent to swamps (not calcareous). The moist rich soil is frequently sandy to some degree. Commonly associated trees are Sugar Maple, Hemlock and Red Maple. Often associated with other locally rare species such as (in Greenock Swamp) Painted Trillium.

DESCRIPTION
Height: Flower stalk up to 60 cm.
Leaves: Two, basal, spreading, up to 20 x 15 cm
Inflorescence: Raceme, ± 20 “watery” flowers.
Flowers: Dorsal sepal up to 8 x 10 mm, rounded. Lateral sepals up to 16 x 9 mm, linear, curved up. Lateral petals up to 13 x 4 mm, linear, spreading. Lip oblong-linear, up to 24 mm. Spur up to 4 cm. **Greenish-white.**
Fruit: Club shaped capsule, stalked.
Flowering: Late June to early August.

50. Great Plains Ladies' Tresses **G4 S3**

***Spiranthes magnicamporum* Shev.** **Orchid Family**

[*speira* — coil, *anthos* — flower (Gr.) *magna* great, *campus* — plain (L.)]

The Ladies' Tresses are an easy group to recognize with their small white flowers coiling round a slender spike. As its Latin name suggests this is a species of the Great Plains of the mid-west and is very much on the eastern edge of its range in Ontario. It is only found in the southwestern corner of the province. In this area there are only three Ladies' Tresses which have more than a few flowers on the spike. However, Hooded Ladies' Tresses (*S. romanzoffiana*) is much smaller. Most difficulty will be found in separating it from Nodding Ladies' Tresses (*S.cernua*) which is about the same size, and in both species the flowers face down — "nod." However, Great Plains Ladies' Tresses tends to be found in drier sites than the other two. If you have a good sense of smell, you will be able to detect a pleasant odor which some say smells like almonds and others say vanilla! However, other species of Ladies' Tresses are also pleasant smelling. In general, it will be found in unshaded areas on basic soil — shorelines, cobbles, dune ridges, dry grasslands, and marshes that dry up in summer.

Local Distribution and Habitat: Very rare. Found only at three open, treeless sites in the south-central Bruce Peninsula, two of which are clearly calcareous; the other was in the Keppel Township unit on the edge of Wiarton. This species has not been found in Grey County.

DESCRIPTION

Height: 10-30 cm.

Leaves: 14 x 1.2 cm, linear-lanceolate, withering early.

Inflorescence: Densely spiralled with several vertical ranks. Floral bracts ovate-attenuate 1-3 cm long.

Flowers: Whitish-cream, fragrant. Sepals linear-lanceolate, 7-11 x 1.5-3 mm. Lateral petals linear, attached to dorsal sepal. Lip oblong-ovate, 7-11 x 3-6 mm. Apex reflexed, margin crisped. Centre fleshy and yellowish (greenish-yellow in Hooded). No spur.

Fruit: Capsule, elliptic 10 x 6 mm.

Flowering: Late September to October.

GRASSES

(Poaceae)
[*Poa* — grass (Gr.)]

Grasses are distinguished from the Sedge family by their hollow, rounded stems with joints where the leaf blades branch off from their sheathing bases. Their flowers are tiny florets with no petals or sepals, just two pairs of chaffy bracts surrounding the three (or 6) stamens and the pistil. The florets are grouped into spikelets which are arranged in various groupings, branched or unbranched, on the flowering stem. Grasses are wind pollinated: the flower stalks rise above the leaves, so that their copious pollen can catch the breeze and be wafted to the feathery stigmas of other plants of the same species. The fruits are single seeded grains. In our area, there are 129 grass species, of which 6 are considered provincially rare. Of these, four described here are associated with sandy coastal areas on Lake Huron including Georgian Bay, especially where there are dunes.

Sand dunes are a very fragile environment dependent for their stability on the vegetation binding them in place. Unfortunately, they are often close to sandy beaches which tend to attract tourists and cottage development. Allowing people to scramble all over dunes and, worse still, ride on them with dirt bikes and A.T.V.'s is disastrous. Removal of the vegetation by this means can not only cause extinctions but also major blowouts. The continued presence of these provincially rare species is an indicator of a healthy natural environment.

51. Beach Grass (Marram Grass, Dune Grass) G5 S3

***Ammophila breviligulata* Fern.**

[*Ammos* — sand; *philos* — beloved (Gr.) *brevi* — short; *ligulata* — ligules]

Beach Grass is an eastern North American endemic which is found on dunes and dry sandy shores around the Great Lakes, Lake Champlain and the Atlantic coast from Newfoundland to North Carolina. It is similar in appearance and habitat to European Marram Grass (*A. arenaria*) a coastal species which in Europe is regarded as very important in maintaining the stability of sand dunes. Beach Grass has a specific habitat requirement of dunes or sandy shores. The sporadic occurrence of this grass in our area is an indication of its need for this environment which can easily be disturbed. This tall grass with flat leaf blades and densely flowered spikes which are about 20 cm (8 inches) long is easily recognized in its typical habitat on shoreline sand dunes. The sharp edges of the leaf blades can easily cut one's hand.

Local Distribution: Rare in this area. Known only along the Lake Huron shore in Bruce County, sporadically, in suitable habitat, from near the Huron County line north to the north end of Sauble Beach, and at Dorcas Bay. Often abundant where present.

Local Habitat: Dry sand dunes, usually unshaded ones at the shoreline, but occasionally very slightly inland in openings on lightly treed dunes.

Herbarium Specimen

DESCRIPTION

Height: 50-100 cm
Leaves: Rolled, ± 5 mm wide, rough and ridged.
Inflorescence: Dense spike, 10-40 x 1-2.5 cm.
Spikelets: One flowered, cylindrical, tightly packed. Bracts mauvish when young, straw coloured at maturity. Florets with a tuft of hairs at base.
Flowering: Summer (mainly).
Fruits: Evident from summer to fall.

52. Sand Reed Grass **G3 S3**

***Calamovilfa longifolia* (Hook.) Scribner var. *magna* Scribner & Merr.**

[*Calamos* — reed, *vilfa* — grass (Gr.); *longi* — long, *folia* — leaf (L.)]

The typical variety of Sand Reed Grass is a quite widespread, mostly western, prairie plant which does not occur in our area. It is found in northwestern Ontario but it is very rare. The variety we have here is mainly an endemic of Lake Superior, Lake Michigan and Lake Huron with a few outlying populations along the Mississippi River in Illinois, Wisconsin and Iowa.

The hard, scaly rhizomes which this plant produces make it an excellent sand binder. Its very localized occurrence is an indication of very precise habitat requirements. The main threats to the continued existence of this rare species are recreational use of dunes and shoreline development. The flowering stems are very striking with the delicate feathery arrangement of the inflorescence. Especially in the early stages, the inflorescence has an unusual purplish-red tinge with white dashes created by the contrast between the light shiny inner bracts and the darker outer bracts of the spikelets. It can almost sparkle in the sun. As with the previous species, it can only be found in sandy areas.

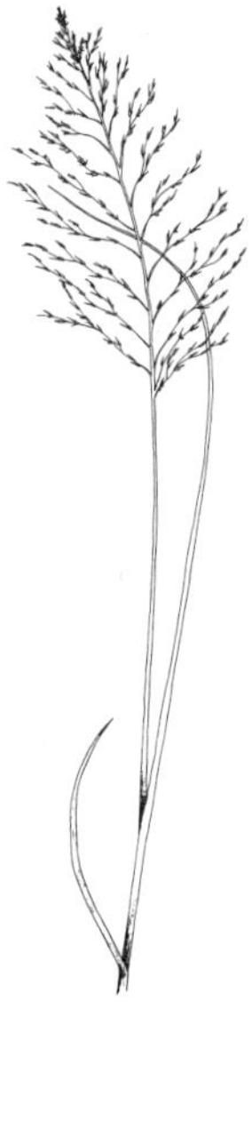

Local Distribution: Very uncommon here but, nevertheless, more common than elsewhere in Ontario. Known only along the Lake Huron shore in Bruce County from near the Huron County line to the southern part of the Oliphant area. Within that narrowly defined strip, it could be considered fairly common. When present, it is often abundant and can be the dominant plant.

Local Habitat: Dry, essentially open sites on sand.

(a) On open shoreline dunes, very often with Beach Grass (51).

(b) Up to 2 km inland, in openings in forests, on old dunes, or on sandy roadsides. These populations are generally small.

(Note: Beach Grass (51) and Great Lakes Wheat Grass (53) are not known at inland sites).

DESCRIPTION

Height: ± 60-100 cm from long rhizomes.

Leaves: Hairy sheaths to 15 cm, blades hairless to 50 x 1.2 cm. Tapering to a long thin point, margins inrolled.

Inflorescence: A large panicle ± 3 times as long as wide.

Spikelets: Single flowered, smooth 5 mm long.

Flowering: Late summer.

Fruits: August, September (October?).

53. Great Lakes Wheat Grass G3 S3 (subspecies)

***Elymus lanceolatus* (Scrib. & Smith) ssp. *psammophilus* (Gillett & Senn.) Löve**

[Synonyms: *Agropyron psammophilus; A. dasystachyum var. psammophilus; Elytrigia dasystachya var. psammophilus*] [*Psammos* — sand; *philos* — beloved (Gr.)]

The confusion caused by this species is evident in the number of name changes. It is now considered to be a separate subspecies of a more widely distributed grass of the western cordillera and northern Great Plains. The other subspecies (*E. lanceolatus* ssp. *lanceolatus*) has not been reported from our area. Great Lakes Wheat Grass is a very restricted endemic, found exclusively on the sand dunes of the shorelines of the upper Great Lakes. It is a very rare in Ontario, found only on Manitoulin Island and in Bruce, Simcoe and Huron Counties. It can be quite abundant locally.

The spikelets in this species have 3-6 flowers, flattened and alternate on the flowering spike. A similar species is Slender Wheat Grass, which is widespread on poor soil in Bruce County. The blue-glaucous appearance of all the vegetative parts of Great Lakes Wheat Grass, plus its hairy spikes make it easy to pick out from other grasses once it is known. The two species can hybridize but this happens only rarely. The very common Couch grass is also related; the running habit and the spike are similar, but it is not hairy or blue-glaucous.

Local Distribution: Rare in our area, found mainly along the Lake Huron shore of Bruce County from near the Huron County line north to a few kilometres before Tobermory, occasionally very slightly inland, recorded from 14 locations. Also known from a bay east of Tobermory. Most prevalent in Huron Township.

Local Habitat: Well drained sand, relatively unshaded, mainly shoreline sand dunes. Often with Sand Reed Grass and Beach Grass.

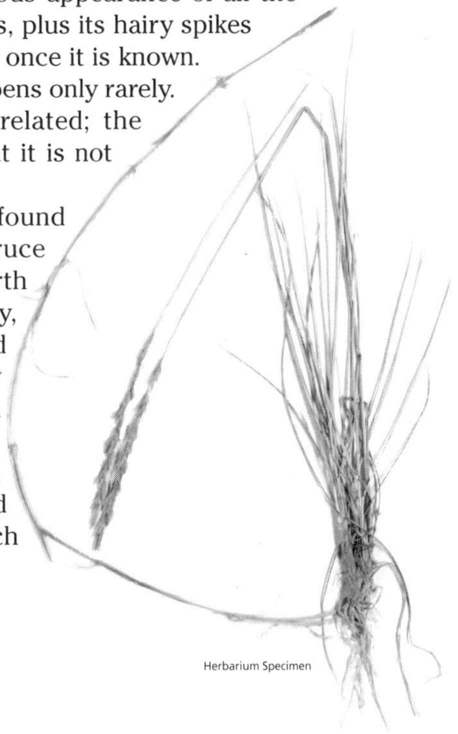

Herbarium Specimen

DESCRIPTION

Height: 30-90 cm from running rhizomes.

Leaves: Blue-glaucous, ± 2 mm wide, rolled in when dry.

Inflorescence: Spike 6-25 cm. Densely hairy.

Spikelets: Flattened, alternate. Bracts densely hairy.

Flowering: Early summer.

Fruits: In summer.

54. Canby's Blue Grass G5 S1

Poa secunda **J.S.Presl.**
[*Poa canbyi* Presl.]

The Blue Grasses, in general, are so named because of their bluish-green colour, often with a whitish bloom. Canby's Blue Grass is yet another species which has caused confusion among botanists, having first been named as *Poa secunda* in one area, and later in another as *Poa canbyi* after William Canby (1831-1904) who discovered it. They are now known to be the same species. The specific name *secunda*, which means "turned to one side," is not very applicable to our population, but it pre-dates *canbyi*. However, Canby's name is retained in the English designation. The distribution of this grass is very odd as it has been found in eastern Quebec and in Ontario only in Bruce County and Thunder Bay District. Other than that its main distribution is from Michigan through Minnesota to the Great Plains and on to the west.

Compared with our other commoner Blue Grasses, this has a very compressed panicle, almost a spike, with short, very erect, branches and shining straw coloured spikelets. The leaf blades are short and narrow and covered with a white, waxy bloom. These characteristics, with the tufted habit of growth and particular habitat, should make this easy to recognize if you are lucky enough to find it.

Local Distribution: Flowerpot Island and Bears Rump Island (Tobermory Islands), known since 1936.

Local Habitat: On cliff ledges and at other rocky, relatively unshaded locations.

DESCRIPTION

Height: 20-90 cm. Tufted, no rhizomes.
Leaves: Basal rosettes. Rolled in. 10-15 cm high, ± 1 mm wide.
Spike ± 5 cm long, ± 1 cm wide.
Spikelets: 2-5 flowers. Bracts rough hairy.
Flowering: Uncertain.

55. Northern (Prairie) Dropseed G5 S3

Sporobolus heterolepis **A.Gray**

[*spora* — seed, *balein* — free grain; *hetero* — different, *lepis* — scale (Gr.)]

The Dropseeds are so called because the grains separate from the chaff and drop to the ground. There are over 100 species world wide. We have five representatives in our area but three of them could not be confused with Northern Dropseed as they have flowers in narrow spikes. *Sporobolus cryptandrus* Sand Dropseed has a panicle with single branches alternating on the main stem and short secondary branches with dense spikelets. In Northern Dropseed the branches of the panicle are whorled and the sparse spikelets droop at the end of each secondary branch. The grains are large and spherical. Sand Dropseed also has conspicuous tufts of white hairs at the junction of the leaf sheath and blade, and it grows on sand. With experience, this species is readily recognizable, even when it has dried up later in the season. Early in the season, it has a characteristic delicate green colour visible at a glance when one drives by an alvar, where it forms distinctive mats or sods. Where sods have not formed at a site, this suggests that it has only recently established itself in that location. While this is mainly a prairie species, it also occurs in Quebec and Connecticut. In this province it has been found in a number of counties etc. across south central Ontario from Manitoulin Island and Bruce County to the Ottawa area, as well as in Kenora District.

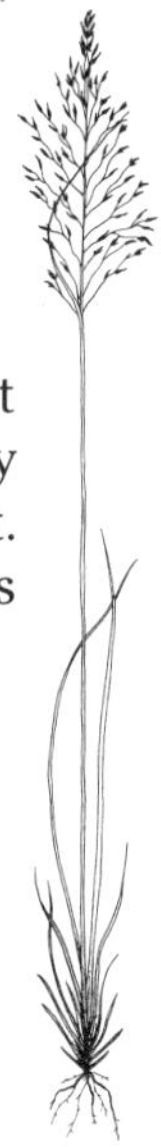

Local Distribution: Northern Dropseed is known only from Bears Rump Island south through the Bruce Peninsula, about equally east and west. The most southern record is from Chiefs Point. On the peninsula it may be said to be uncommon rather than rare, and it is often abundant where it does occur. The Cape Croker prairie population, occupying a few square kilometres, must be one of the largest (if not the largest) populations east of Illinois. The only other part of Ontario where it may be at least as common is Manitoulin District. It seems to have increased significantly on the Bruce Peninsula in the 1970's and 1980's. It was found on Cape Croker as far back as 1905, but even in the 1930's nowhere else. This is in spite of the fact that St. Edmunds and Lindsay Townships, where the majority of the present day populations are to be found, had been heavily botanized for decades.

Habitat: In this area it most often occurs in dry to mesic locations (sometimes seasonally wet) in alvars. These are dolostone, except for the Cape Croker prairie alvars which are on limestone and calcareous shale. Sometimes found on bedrock shores of Lake Huron; rarely on sand under very light shade of pines.

DESCRIPTION

Height: 40-100 cm. Tightly clustered. Forming mats.
Leaves: Long, narrow, rolled in. Pale Green.
Inflorescence: Panicle. Branches whorled, ascending.
Spikelets: 3-6 mm, one per branchlet.
Flowering: Summer. Fruits August to September.

56. Porcupine Grass G5 S3

Stipa spartea **Trin.**

[*stype* — tow. *spartea* — coarse rushy grass (Gr.)]

This is one of the easiest grasses to recognize. One of the dark brown chaffy bracts around the grain produces a long awn, like a porcupine quill. This bunch of "quills" reminded Linnaeus, who named this genus, of a bunch of flax or "tow" ready for spinning — porcupines not being a familiar sight in Sweden! His compatriot Karl Trinius named this particular species *spartea* in 1831, when the botanical exploration of North America was at its height. Like the previous two species, this is a prairie plant, generally found on sandy soils. The destruction of its native habitat has made it a rare or endangered species in a number of States and it is very much on the edge of its range in Ontario. It is found in several counties in southwestern Ontario and on Manitoulin Island where there are prairie affinities. Similarly, it occurs in Kenora and Rainy River Districts. However, it is also reported from Algoma District and Simcoe County. Its presence in Bruce and the last two areas is harder to explain. It could be a relic of past climates. In Ontario it shows some tendency to occur along Great Lakes shores.

Local Distribution: Extremely rare. Known from only one location — a bay to the east of Tobermory. It was recorded as far back as 1933 and was still present recently. In the 1930's only a few plants were present. In recent decades the population has become somewhat larger.

Local Habitat: On sand dunes above the water line, partly in very thin shade from trees.

DESCRIPTION

Height: ± 60-100 cm. In tufts. No rhizomes.

Leaves: Blades 2-3 mm wide. Smooth below, rough with short hairs above.

Inflorescence: Panicle 10-20 cm, narrow, nodding with few branches.

Spikelets: Mostly one per branch. Dark brown bract with long awn — 12-20 cm, stiff, twisted at base.

Flowering: Uncertain.

57. Hill's Pondweed

G3 SC S2 THR

***Potamogeton hillii* Morong**

Pondweed Family

[*Potamos* — river; *geton* — neighbour (Gr.)]

The name of this family, indicates its close relationship with water, and dates back to Pliny. It was adopted as the generic name by Linnaeus in 1753 and these species are now considered sufficiently distinct to be placed in a separate family. There is only one genus with 100 species worldwide. They are basically freshwater plants, although a few may be found in brackish water near the sea. The water is usually fairly shallow as they are rooted to the bottom with rhizomes. Some species produce leaves above the water but many are completely submerged. The leaves are undivided, although the shapes vary. The flowers are usually on small spikes which, in most cases, rise above the surface. Nineteen species have been reported from Bruce and Grey Counties, only one of which is non-native. Although there are so many species, the presence or absence of one of the rarer ones can be a good indicator of the health of the environment. Hill's Pondweed was named after E.J.Hill who was the first to collect it in Michigan in 1880. It occurs in a restricted area of eastern North America from Vermont and Massachusets to Pennsylvania, Ohio and Michigan. Ontario is the only Canadian province in which it is found. It grows completely submerged with slender, branching stems which may be up to one metre long, and have long narrow leaves. It has only one or two whorls of tiny flowers and the achenes are rounded with three keels. It can be confused with the much commoner Leafy Pondweed (*P. foliosus*). The latter is less robust and has concave sided fruits with one wing-like keel. Hill's is our only pondweed with leaves more than 1 mm wide, that have three veins and a bristle-tipped apex.

Local Distribution: While it is very uncommon overall in this area, it is probably less rare than in most of Ontario. It is most prevalent in the northern half of the Bruce Peninsula, but it may be found widely scattered throughout Bruce County. It is genuinely rare in Grey County. It probably occurs very sparingly in some township units other than those for which it is mapped, as it is easily overlooked. It has probably been increasing in recent decades, at least on the Bruce Peninsula, from which it was reported as early as 1901.

Local Habitat: Small shallow bodies of water, often calcareous, such as sluggish creeks, brooklets, ditches, small ponds, and, in at least one case, in the edge of a lake. Frequently at sites shaded by trees.

DESCRIPTION

Stems: up to 1 m. Branching.

Leaves: All submersed. Linear, acute, 30-70 x 1-2 mm, often with a bristle tip. Three lengthwise veins. Stipules whitish.

Inflorescence: Short stalked spike ± 10 mm. 1(or 2) whorls. 3-6 flowers.

Flowers: Tiny with 4 clawed tepals, 4 stamens, 4 ovaries.

Fruit: Achene, 2.5-4 mm, 1 sharp dorsal keel and 2 lateral keels. It floats for dispersal.

Flowering: Uncertain.

58. Branching Bur Reed **G4G5 S1**

***Sparganium androcladum* (Eng.) Morong** **Bur Reed Family — Sparganiaceae**

[*sparganion* — swaddling band, *andros* — male, *clados* — branch (Gr.)

The Latin name of the Bur Reeds is an ancient one dating back to classical times. It is derived from the word for swathing band and refers to the flexible strap-like leaves. The spherical heads of spiky flowers and fruits are reminiscent of burs but they do not have hooks. In this species, the female (pistillate) flower heads occur on the main axis, and the male (staminate) flower heads above on the main axis or on branches. The specific name is derived from this habit. Six species of Bur Reed, not including *Sparganium fluctuans* which is a historical record, occur in this area. Giant Bur Reed is much bigger, Green Fruited Bur Reed is smaller and has green achenes, Least Bur Reed is smaller and the stems float. Narrow Leaved Bur Reed is only found on the Bruce Peninsula and is very rare. This leaves American Bur Reed which is very similar to Branching Bur Reed, except that its achenes are dull brown, whereas in Branching Bur Reed they are shining, pale brown to straw coloured. These are all plants of muddy shores and shallow water of ponds, lake margins and marshes. This species has an unusual range being found in New England, and as far south as Virginia, and also around the lower Great Lakes. In the west it is found from northern Minnesota to Missouri with a gap between the two areas south of Lake Michigan. In many of the states where it is found it is endangered and in Ontario it is recently reported only from southern Grey County.

Local Distribution: Extremely rare. Known only at Eugenia Lake, and at Spring (Rattlesnake) Creek (historical) in the northern Bruce Peninsula.

Local Habitat: Marshy bay of lake margin; shallow water elsewhere.

Herbarium Specimen

DESCRIPTION

Height: 40-100 cm. Stout, erect triangular stems.

Leaves: ± 80 cm x 5-12 mm.

Inflorescence: Globose, stalkless cluster. Pistillate heads below on main stem. Staminate heads above or on branches.

Flowers: No sepals or petals, a few chaffy scales. Pistillate flowers with a single celled ovary. Stamens mostly 5.

Fruits: Achene, spindle-shaped with a curved beak. Shining, pale brown to straw coloured.

Flowering: Uncertain.

Other Species of Monocotyledons Listed as Rare in this Area by NHIC

Historical — pre 1975:

Long Scaled Tussock Sedge G5 S2

(*Carex haydenii* Dewey)

A specimen was collected in 1934 at Gould Lake in Amabel Township. There is a possibility of misidentification.

Bluntscale Bulrush G5? S2?

(*Scirpus smithii* A.Gray)

This was collected once from one location in southern Grey County.

Greene's Rush G5 S3

(*Juncus greenei* Oakes & Tuckerman)

This species was collected along the shore of the Saugeen Indian Reserve in 1901, but has not been seen since.

Small White Lady's Slipper G4 END S1 END

(*Cypripedium candidum* Muhl. ex Willd.)

This is an endangered species both provincially and nationally, but the single record for this area is from near Port Elgin and dates back to 1903.

Western Smooth Brome G? SH

(*Bromus inermis* Leys. ssp. *pumpellianus* (Scrib.) Wagn.)

This species was collected from one location near Tobermory in 1933 and there in no record of any sightings since 1934.

Species not native to this area, "not definitely S1-S3" or unconfirmed:

Silver Sedge G5 S2S3

(*Carex argyrantha* Tuckerman)

This species has been reported as occurring, very rarely, on the Bruce Peninsula in Bruce County. It may have been misidentified.

Field Sedge G5 S3

(*Carex conoidea* Schk. ex Willd.)

This species was collected in a ditch near Sauble Beach in the 1980s. It may be adventive, rather than native.

Beautiful Sedge G4 S3S4

(*Carex formosa* Dewey)

This species has been reported in Grey County from the South Saugeen River ANSI and the Beaver Valley Lowlands ANSI. It is very rare in this area.

White Fringed Orchis G4G5 S3S4

(*Platanthera blephariglottis* (Willd.)Lindl.)

This species is known from several townships in Bruce and Grey Counties off the peninsula. It is rare overall in our area.

Forked Panic Grass G5 S2

(*Panicum dichotomum* L.)

A specimen was collected on the Meaford Tank Range in 1994, but there is a possibility of misidentification.

Rough or Tall Dropseed G5 S1S2

(*Sporobolus asper* (Michaux) Kunth)

Collected from roadsides at one location in Grey County and near Southampton in Bruce County. In both cases, presumed to be adventive.

Melic Oat Grass G4 S3S4

(*Trisetum melicoides* (Michaux) Vasey)

This species is found on the Bruce Peninsula where it is uncommon. It also occurs rarely in southern Bruce County. Not known from Grey County.

Plant References

Argus, G.W. et al. 1982-1987. **Atlas of the Rare Vascular Plants of Ontario.** National Museum of Natural Sciences.

Argus, G.W. & K. Pryer. 1990. **Rare Vascular Plants in Canada.** Can. Mus. of Nature.

Britton, N.L. & A. Brown. 1913. **An Illustrated Flora of the Northern United States and Canada.** Dover Publications Reprint, New York.

Bruce-Grey Plant Committee. 1997. **A Checklist of Vascular Plants for Bruce and Grey Counties. 2nd. Ed.**

Bruce-Grey Plant Committee. 1997. **The Orchids of Bruce and Grey.**

Bruce-Grey Plant Committee. 2000. **The Asters, Goldenrods and Fleabanes of Grey and Bruce Counties.**

Budd, A.C. & K.F. Best, 1964. **Wild Plants of the Canadian Prairies.** Canada Dept. of Agriculture, Ottawa.

Dore, W.G. & J. McNeill. 1980. **Grasses of Ontario.** Research Branch, Agriculture Canada.

Gleason, H.A. 1952. **The New Britton & Brown Illustrated Flora of the Northeastern United States and Adjacent Canada. Vols. 1, 2 & 3**. Hafner Press, New York.

Gleason, H. & A. Cronquist. 1991. **The Manual of Vascular Plants of the Northeastern United States and Canada.** The New York Botanical Garden.

Grieve, M. 1980. **A Modern Herbal,** Penguin Books, London.

Holmgren, Noel. 1998. **The Illustrated Companion to Gleason & Cronquist's Manual.** New York Botanical Garden.

Johnson, J. 1990. **The Vascular Flora of Three Regions Comprising Bruce and Grey Counties, Ontario — with Emphasis on Rare Taxa.** Ontario Min. of Natural Resources.

Kershaw, L. et al. **Plants of the Rocky Mountains**. Lone Pine Publishing. Edmonton, Alberta.

Kozloff, E.N. 1976. **Plants and Animals of the Pacific Northwest.** J.J. Douglas Ltd. Vancouver, B.C.

Krotkov, P.V. 1940. **Botanical Explorations in the Bruce Peninsula, Ontario.** Trans. Royal. Canadian Institute. No. 9. Vol. 23, Part 1.

Lakela, Olga. 1965. **A Flora of Northeastern Minnesota**. University of Minnesota

Montgomery, F.H. 1966. **Plants From Sea to Sea.** The Ryerson Press. Toronto.

Morton, J.K. & J.M.Venn. 1984. **The Flora of Manitoulin Island**. Biology Dept. University of Waterloo, Ontario.

Morton, J.K. & J.M.Venn. 1987. **The Flora of the Tobermory Islands.** Univ. of Waterloo.

Morton, J.K. & J.M.Venn. 1990. **A Checklist of the Flora of Ontario.** University of Waterloo.

Newcomb, L. 1977. **Newcomb's Wildflower Guide**. Little, Brown & Co., Boston.

Newmaster, S.G. et al. 1998. **Ontario Plant List**. Ont. Forest Research Inst., Sault Ste. Marie.

Oldham, M. 1999. **List of Rare Vascular Plants of Ontario.** N.H.I.C Peterborough, ON.

Riley, J.L., J.Javala, S.Varga. 1996. **Ecological Survey of the Niagara Escarpment Biosphere Reserve. Vol I & II.** Ontario. Ministry of Natural Resources.

Scoggan, H.J. 1955. **Flora of Manitoba**. National Museum of Canada. Bull. No. 140.

Scoggan, H.J. 1978, 1979. **The Flora of Canada, Parts 2,3,4**. National Museum of Canada.

Semple, J. et al. 1996. **The Asters of Ontario**. Revised ed. Univ. of Waterloo.

Semple, J. et al. 1999. **Goldenrods of Ontario**. 3rd ed. University of Waterloo.

Soper, J.H. and M.L. Heimburger, 1982. **Shrubs of Ontario.** Royal Ontario Museum, Toronto.

Stearn, William. T. 1983. **Botanical Latin.** 3rd. Edition. David & Charles Publishers, Newton Abbott, Devon, U.K.

Sugden, Andrew. 1984. **Longman Illustrated Dictionary of Botany.** Longman Group, U.K. Ltd. Harlow.

Voss, E.G. 1972, 1985, 1996. **Michigan Flora. Vols. 1, 2 & 3.** Univ. of Mich., Ann Arbor.

The Flora of North America. Volume I & II, 1993, Volume III 1997, Volume XXII, 2000. Oxford University Press.

The Compact Edition of the Oxford English Dictionary 1971. Oxford University Press.

1. Butternut G3G4 S3

***Juglans cinerea* L.**

Butternut (or White Walnut) trees were a common sight in deciduous forests in Grey and Bruce Counties, but in recent years they have been attacked by a fungal disease known as Butternut Canker with such devastating results that their future is becoming questionable. The fungus grows in the cambium or inner bark of the tree, eventually girdling it and destroying its ability to conduct water to the crown. It also breaks down the bark allowing entry to other fungal enemies. The presence of Butternut Canker is indicated by sooty black patches from which an inky fluid oozes in the spring. This tree is a species confined to temperate eastern North America. The disease first appeared in the United States around 1950 and has spread so rapidly that many states are now almost devoid of Butternuts.

Though far from abundant, Butternut is probably more common in this area, especially in Grey County and the southern portion of the Bruce Peninsula, than in most parts of Ontario. The fairly close proximity of these trees has facilitated the spread of the disease, so that today the majority of standing Butternuts are diseased or already dead.

In recent decades the possibility of Dutch Elm disease causing the extinction of the North American elm species was a concern. However, although greatly reduced in number, elms have stabililized and seem to be increasing again. The American Chestnut has also been almost extirpated by Chestnut blight which came from Asia or Europe, where the trees have a natural resistance to it. Chestnuts tend to form a clone of sprouts from the roots when the tree dies, so something survives and they continue to persist in many localities, but usually they do not live to maturity. There are still tiny scattered populations throughout its original range and there is hope that it, too, will eventually recover. Butternuts do not usually root sprout. Once the tree dies it is eliminated from that site. This reduces the possibility of the species developing a natural immunity to the disease.

2. The Monarch Butterfly **SC S5**

***Danaus plexippus* L.**

The Monarch is one member of a family of 300 species; large, brightly coloured, and mostly tropical, whose caterpillars feed on milkweeds (sometimes also on the related Dogbane). Milkweeds are so-called because they produce a milky kind of latex which oozes out of broken stems. This is poisonous to many animals, which makes it a useful survival characteristic for the plant. The Milkweed Caterpillars have evolved a resistance to this poisonous substance, so they can exploit it as a food source and a protection against predation. In Bruce and Grey Counties we have three species of milkweed, including the Common Milkweed (*Asclepias syriaca* L.) whose fat pods, bursting with fluffy seeds, are a familiar sight along roadsides. Unfortunately, this is recognised as a weed in Canada because of its toxicity to livestock, which means farmers do their best to destroy it. To further complicate their lives, Monarch Butterflies are migratory. It was not until the 1950s that Fred and Nora Urquart of the University of Toronto started tagging Monarchs. It took more than twenty years to track them across the United States to their overwintering sites in the mountains west of Mexico City. During the summer they produce a few short-lived generations, but the last set of adults to emerge in September fly south across the United States to central Mexico, where they overwinter in their millions in forested sites. Towards the end of winter the butterflies start to mate and move down the mountains. In March they travel north and east, and lay their eggs on milkweed plants they encounter in the southern states. Then they die. The stripy caterpillars hatch out and, within about a month, they pupate and emerge as adults which fly further north before they lay their eggs. After a few generations, they arrive in the northern states and Canada in late June, with time to produce more broods before the last long-lived generation completes the cycle by flying off to Mexico. Needless to say, such a complicated pattern which has evolved over countless millenia is very easy to interrupt, with possible fatal consequences. Therefore, even though their current rarity rating is S5, COSEWIC at one time designated this species "Vulnerable." There has been much concern about conservation of the overwintering areas which are threatened by logging and human population pressures. On November 9, 2000, the Mexican president, Ernesto Zedillo, announced the creation of the Monarch Butterfly Biosphere Reserve linking existing reserves and giving protection in 59 259 hectares. The World Wildlife Fund announced at the same time that it is working on measures to control pesticide use at migration stopovers. Unfortunately, the latest news in March 2001 is that there has been a severe winter die-off, which has been attributed by some to unusually severe weather and by others to spraying with DDT — the latter substance having been long banned in North America and Europe, but still marketed extensively by international chemical companies in less wealthy countries.

Appendix I: The Distribution of Species in Grey and Bruce Counties

On these maps the square is marked if that species has been found **at least once** in that township unit. This gives no indication of the rarity or abundance of any species. The townships were chosen originally, rather than a more systematic grid, because some data were available. There is variation in the size of the townships but only Sarawak is markedly smaller. For the purposes of this study, the boundaries of the townships adjacent to the towns and villages have been extended through them (for adminstrative purposes the city, towns and villages were separate from the townships). For this reason we refer to the divisions on the map as "township units." Larger communities presented some problems and they have been dealt with as follows:

1. Wiarton — from Highway 6 (Berford Street) east to the bay and south of William Street (traffic signal) is included in Keppel. The remainder is in Amabel.
2. Owen Sound — Mostly in Derby. The boundary of adjacent Sydenham is defined by Highway 6/10 extended north along 9th Avenue East to 32nd Street and west down to the bay. The Derby-Sarawak Township line is extended to the Pottawatomi River and along it to the bay; north of this line is in Sarawak.
3. Those municipalities on the southern county boundaries, e.g. Lucknow and Mount Forest, have only the portion north of the streets aligned with the boundary included. Similarly, where more than one township is involved, the township lines would be extended along streets closest to the dividing line.
4. Tiverton, Hanover and Durham etc. are divided along the roads aligned with the township boundaries.
5. Records from the First Nations Hunting Reserves 60A and 60B are included in St. Edmunds. The Cape Croker First Nation 27 is included in Albemarle and Saugeen First Nations 28 and 29 are included in Amabel.
6. The boundary between the Saugeen and Amabel Township units cuts through Southampton. Going west, at the west end of Chippewa Hill, it leaves the south boundary of Saugeen First Nation 29 and follows provincial highway 21 west to the Saugeen River bridge and then west along the river to its mouth.
7. First Nation 1 is comprised of some of the islands off Oliphant and Red Bay. Burke Island and those to the south are included in Amabel. The islands north of Burke Island are included in Albemarle.

It should be noted that municipalities are in a state of flux at this time with many amalgamations taking place, but the roads bordering the original townships will remain. The only revision we have made in this respect is to change the name of Collingwood Township and Thornbury-Clarksburg to The Town of The Blue Mountains.

Map of Township Units Based on Historical Townships

Glossary

Achene A one-seeded fruit which does not split open (indehiscent).

Adventive Exotic — not firmly established in a new environment.

Alvar Open limestone or dolostone pavement.

Amphi-Atlantic Found on both sides of the Atlantic Ocean.

ANSI Area of Natural and Scientific Interest.

Awn Stiff needle-like extension of a part of a plant, especially grass bracts.

Bract A small leafy structure associated with a flower or inflorescence. In grasses these are given special names such as glume, palea and lemma. In this book the use of too many technical terms has been avoided so they have been called bracts.

Calciphile A plant found mostly on substrates with a high calcium content.

Calyx Collective name for the sepals — outer ring of the flower.

Cere Naked wax-like area at the base of a bird's beak (in a raptor).

Clone A population of plants all with the same genetic constitution.

Cordate Heart shaped.

Corm The swollen base of a stem with food stored in it, forming a reproductive structure which can remain dormant through the winter.

Corolla Collective term for petals.

Corymb See Inflorescence.

Covert Small feathers on top of the wings and over the tail feathers.

Cyme See Inflorescence.

Discoid Resembling a disc. Flower head in the Aster family without "petals" (ligules).

Dolostone Composed largely of dolomite (Magnesium calcium carbonate — $MgCa(CO_3)_2$). Slightly harder and more resistant to weathering than limestone. Dolostone is the common surface rock along the Niagara Escarpment.

Driftless Area An area with no glacial deposits.

Dune-panne Complex Sand dunes alternating with low-lying wet areas.

Elliptic See Leaf Shapes.

Endemic A species which is confined to a particular, well defined geographic area.

Extirpation Elimination of a species in a particular area, not necessarily total extinction.

Fauna Species belonging to the animal kingdom e.g. birds, mammals, insects etc.

Fen An open area with slowly moving water and low vegetation growing on a non-acid substrate. In some cases vegetation forms a floating mat.

Flora Species belonging to the plant kingdom.

Frond The leaf of a fern.

Glaucous Covered with a fine, waxy powder (bloom) giving a bluish or whitish cast to the surface of a leaf or fruit.

Hybridize Species which may appear to be somewhat different but which are closely related genetically. They may cross when they come into contact in marginal habitats. The intermediate individuals thus produced are known as hybrids.

Inflorescence The arrangement of individual flowers on a flowering stem.

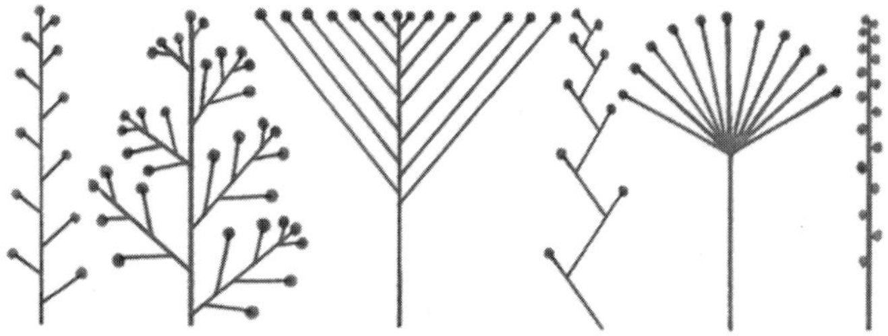

Raceme **Panicle** **Corymb** **Cyme** **Umbel** **Spike**

Indigenous A species native to the area under consideration.

Keel A ridge like the keel of a boat e.g. on a fruit.

Lanceolate See Leaf Shapes.

Leaflet One section of a divided leaf, such as a fern frond.

Leaf Shapes:

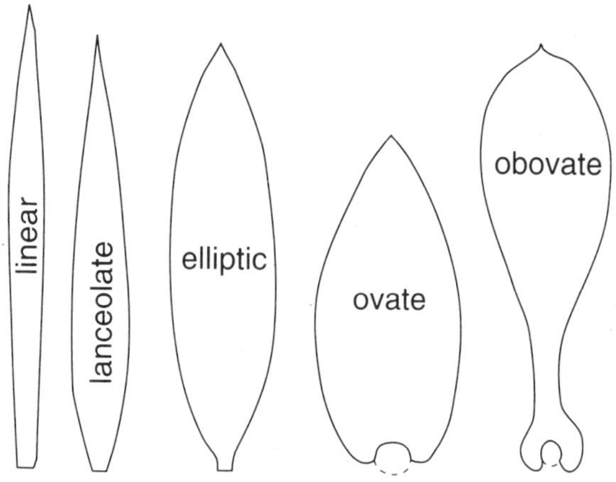

Linear See Leaf Shapes.

Ligule A small leafy appendage arising where a grass leaf spreads out from the sheath into the blade.

Margin Edge of leaf.

Marl A soil or earthy deposit consisting of sand and/or clay with a very high Calcium carbonate content (high pH). Often found in shallow lakes or fens.

Mesic Moderately moist, neither excessively wet nor dry.

Niagara Escarpment The geological formation which runs through southern Ontario from Niagara to Tobermory and on through Manitoulin Island. It is the rim of an ancient tropical sea and is often capped with fossil coral. Where exposed, as it is through most of Bruce and Grey Counties, it forms a steep ridge, in this area often with cliffs facing Georgian Bay and a gentler slope southwestwards.

Non-cyclical Decline A decline in population which is not due to the natural rise and fall of population numbers which occurs under natural conditions.

Obovate See Leaf Shapes.

Ovate See Leaf Shapes.

Palmate Of leaflets arranged in a fan-shaped pattern like the fingers of a hand.

Panicle See Inflorescence.

Petiole Leaf stalk.

Pinnate Of leaflets arranged on both sides of a midrib. Leaflets may be opposite or alternate.

Pistillate Of female flowers with ovary, style and stigma (Pistil).

Pome A fruit like an apple in which the seed box is incorporated in the swollen receptacle.

Primary A large flight feather of the bird's wing.

Raceme See Inflorescence.

Rhizome Underground stem.

Riffle A shoal or rocky obstruction beneath the surface of a stream.

Rosette A circular group of leaves, close to the ground, from which the flowering stems arise.

Saprophytic Extracting nourishment from dead organic material.

Sessile Stalkless.

Species A distinct type of plant or animal which will usually not interbreed with another species and produce fertile offspring.

Spike See Inflorescence.

Sporangium The structure on a fern frond in which the spores form.

Spur A hollow extension of part of a flower — often a petal. Usually containing nectar.

Staminate Of male flowers with stamens.

Stigma The tip of the style — usually sticky, on which pollen is deposited to fertilize the egg cell(s) in the ovary, from which the seed(s) develops.

Stipules A pair of small leafy appendages at the base of a leaf stalk.

Style The structure arising from the ovary and supporting the stigma.

Subleaflet A subdivision of a leaflet.

Subspecies A distinct group within a species which has characteristics that make it significantly different from other groups within the species. Often geographically separated or with different ecological requirements, but still able to interbreed with other subspecies and produce fertile offspring.

Taxon A distinct type of plant or animal. This term applies to a species, a subspecies or a variety. (Plural: taxa).

Tepal Used when there is a single whorl in the flower, which is not clearly composed of sepals or petals.

Umbel See Inflorescence.

Variety A type of plant within a species or subspecies with some distinct inheritable differences.

Whorl The leaves or flowers arising in a circle from a common point on a stem.

Index

ANIMALS

PLANTS

Notes

COUNTIES OF GREY AND BRUCE
REPORT OF NEW OR RARE SPECIES

Name of species: Latin: ______________________

English: ______________________

Name of observer: ______________________

Address: ______________________

______________________ e-mail ______________________

Telephone number: ______________ Fax Number ______________

Name(s) of other observers: ______________________

Date of observation: ______________________

Number observed? ______________________

Do you have a photograph? ________

On what basis was the species identified?

Location of species: ______________________

County: ______________ Township: ______________

Lot and concession numbers: ______________________

Who owns the property? ______________________

Precise directions to location (include a sketch map, if possible):

If available give: 1:50 000 Topographic map number: ______________

UTM 6 digit grid # & square: ______________________

Latitude: ________ Longitude: ________

Would you be able to take someone to view the location? ______________

Any other information?

This form may be copied.

COVER PHOTOGRAPHS AND PHOTOGRAPHERS

Front Cover:

Eastern Prairie White Fringed Orchid (*Platanthera leucophaea*), Nelson Maher
Peregrine Falcon (*Falco peregrinus*), George Peck
Spotted Turtle (*Clemmys guttata*), Donald R. Gunn
Eastern Massasauga Rattlesnake (*Sistrurus catenatus catenatus*), Willy Waterton
Ram's Head Lady's Slipper (*Cypripedium arietinum*), Nelson Maher.

Back Cover:

Bald Eagle (*Haliaeetus leucocephalus*), George Peck
Lakeside Daisy (*Hymenoxys herbacea*), Nelson Maher
Queen Snake (*Regina septemvittata*), George Peck
Wall Rue (*Asplenium ruta-muraria*), Nelson Maher.